Time on TV

This book is part of the Peter Lang Media and Communication list.
Every volume is peer reviewed and meets
the highest quality standards for content and production.

PETER LANG
New York • Washington, D.C./Baltimore • Bern
Frankfurt • Berlin • Brussels • Vienna • Oxford

Paul Booth

Time on TV

Temporal Displacement and Mashup Television

PETER LANG
New York • Washington, D.C./Baltimore • Bern
Frankfurt • Berlin • Brussels • Vienna • Oxford

Library of Congress Cataloging-in-Publication Data

Booth, Paul.
Time on TV: temporal displacement and mashup television / Paul Booth.
p. cm.
Includes bibliographical references and index.
1. Time on television. I. Title.
PN1992.8.T56B66 791.45′6—dc23 2012005683
ISBN 978-1-4331-1570-7 (hardcover)
ISBN 978-1-4331-1569-1 (paperback)
ISBN 978-1-4539-0771-9 (e-book)

Bibliographic information published by **Die Deutsche Nationalbibliothek.**
Die Deutsche Nationalbibliothek lists this publication in the "Deutsche Nationalbibliografie"; detailed bibliographic data is available on the Internet at http://dnb.d-nb.de/.

Cover design by Sarah McNabb / FamousAfterIDie.com
Author photograph by Kate Booth

29 Broadway, 18th floor, New York, NY 10006
www.peterlang.com

To Kate

Table of Contents

Acknowledgments

I truly could not have completed this book without the help of many fine individuals, to all of whom I owe a tremendous debt of gratitude.

First, thanks are due to all the anonymous reviewers for their excellent suggestions and ideas as I've revised chapters of this book. Working with Peter Lang has been a pleasure from beginning to end. Mary Savigar, Sophie Appel, Catherine Tung, and the whole Peter Lang editorial team have been a dream to work with.

I also want to acknowledge the following journals and books, in which some previously published work appeared:

Throughout the book, I have used portions of my 2010 article, "Memories, Temporalities, Fictions: Temporal Displacement in Contemporary Television." The final, definitive version of this paper was published in *Television and New Media*, 12/4, Dec 2010 SAGE Publications Ltd./SAGE Publications, Inc., all rights reserved ©, DOI: 10.1177/1527476410392806. Portions of chapters Three and Six have been presented at the "What Is Television?" conference in Portland, OR in Mar 2012, and published in 2012 in *Communication Studies* 63 as "The Television Social Network: Exploring TV Characters" (used with kind permission by Taylor & Francis, New York, http://www.tandfonline.com/). Chapter Four includes portions from my chapter "Effecting the Cause: Time Travel Narratives and *Doctor Who*," first published in the book *Doctor Who in Time and Space*, edited by Gillian Leitch, for McFarland Press. Chapters Four and Five contain some elements from my 2008 article, "Intermediality in Film and Internet:

Donnie Darko and Issues of Narrative Substantiality" from the *Journal of Narrative Theory* 38(3), pp. 398-415. Chapter Five houses material that first found the light of day in "Time and Relative Dimensions On Line: *Doctor Who*, Wikis, and the Production of Narrative/History" in *Interactions* 1(3): 2009, pp. 331–349, © Intellect. Chapter Seven contains some bits of my 2008 article, "Mediating New Technology: The Realization of a Digital Intellect," first published in *Nebula* 5(1), pp. 28–43 and some material from my 2011 article for *Transformative Works and Cultures* 10 titled "Mashups as Temporal Amalgamations" (also presented at the 2011 Cultural Studies Association meeting in Chicago).

Thanks also to all those that have helped shape my thinking about temporal displacement. Greg Bianchini was inspirational at the onset, noting to me once a few years ago that "wasn't it weird that *Lost* and *How I Met Your Mother* had similar storytelling styles?" Further, the thoughtful discussions and helpful comments from the students in my classes at DePaul University helped solidify my thinking about long-form television. Thanks are due to all my students in both of the MCS 351: Topics in Television classes I've taught, "Time Travel on Television" and "Long-Form Narrative Television." Special thanks also go to my grad students in these classes, as well as the students in my MCS 520: Fandom class, my MCS 520: Monsters class, and my CMNS 545: Communciation and Technology class, for their exquisite work and engaging dialogue about *Doctor Who* and other temporally interesting shows. Michael Lachney also offered help and criticism on revisions and ideas.

DePaul University continues to be a wonderful home. I am indebted to my colleagues Luisela Alvaray, Kim Clark, Blair Davis, Michael DeAngelis, and Kelly Kessler for their help and advice during the writing of this book. A special "thank you" goes out to Michael, whose impressive editorial skills vastly improved the introduction of this book. Jacqueline Taylor is as supportive a Dean as one could ask for, and I'm thankful for her continued enthusiasm for my work. I am grateful to the College of Communication, whose 2011 Summer Research Grant helped me complete this book.

My family has been phenomenal throughout the writing of this book. I am so appreciative of the support of my parents, Deb and Colin, and my sister Anna. We may have wildly different taste in television, but at least we agree some of the time (*Gilmore Girls* is great!). My in-laws, Wendy Vincent and Thomas Vincent, have been supportive and caring for as long as I've known them. Extra thanks to Tom for his keen eye and editorial help.

The equal parts love and disdain that can only be achieved by a cat I have experienced times three, as Gizmo, Dusty, and Black Kitty occasionally calmed my nerves with well-timed purrs and belly rubs. This book (and author) also benefited from the unconditional emotional support provided by my stalwart companions, Slinky and Rosie, who always know the best time to take a nap break (1–3 pm).

Finally, very special thanks to my wife, Kate, who has patiently listened to every twist and turn of thought during the writing of this book, and who, at critical moments, could provide a boost of motivation or a thoughtful turn of phrase. Kate is my best critic and my best friend, and I'm grateful beyond words. She is the best editor I know, and has an innate sense of when something doesn't make any.

Section I: Time on TV

In this introductory section, I demonstrate the intimate relationship between television and digital media by introducing the ideas of "Temporal Displacement," "Mashup Television," and "Transgenic Media." I define *Temporal displacement* as shifts in the representation of time. *Mashup television* describes the way that contemporary television integrates new media characteristics with traditional content. By expanding on the notion of temporal displacement both as it appears on mashup television and in online media, I show that the contemporary television form plays with depictions of time as a heuristic device, to mirror user experiences of online, social, and interactive media. This *Transgenic media* represents the integration of new media aesthetics with more traditional media forms. The past decade has seen an increase in the type, frequency, and visibility of television narratives that utilize "time" in unusual and non-traditional ways. Specifically, I claim that *recent mashup television content focuses on shifting temporal expectations in television narrative and on altering representations of time, in order to reflect and teach about changes brought to our everyday life through the use of transgenic media.*

The introduction offers an explanation of this claim in more detail, concentrating both on the content of contemporary television and the temporal mechanics of online media. Through a detailed and critical discussion of the literature and theory behind temporal displacement on television, I examine in greater depth some of the terms I will use throughout this book and offer a practical demonstration of mashup television, temporal displacement, and transgenic media. Such a discussion provides a foundation upon which Chapter One elaborates, through a detailed discussion of the relationship between television and transgenic media.

Introduction

Temporal Displacement and Mashup Television

> *What I do have control over is my own intention. In the space between free will and determinism are these imperceptible gaps, these lacunae, the volitional interstices, the holes and the nodes, the material and the aether, the something and nothing that, at once, separate and bind the moments together, the story together, my actions together, and it's in these gaps, in these pauses where the fictional science breaks down, where neither the science nor the fiction can penetrate, where the fiction that we call the present moment exists.*[1]
>
> —Yu, How to Live Safely in a Science Fictional Universe, 217–18

Perhaps, then, I should start with a conclusion. At the end of his novel *How to Live Safely in a Science Fictional Universe*, Charles Yu concludes that we live our lives in perpetual yet imperceptible gaps: between past and future, between memory and anticipation, between "is" and "was." Our present is always *now*, a fleeting moment that changes as soon as it is experienced. Yu's protagonist, also named Charles Yu, spends ten years in one of these "gaps," these "pauses in time," as he tries to come to terms with the impossible paradox of time travel.

But why open a book about temporality on television with a quote from the conclusion of a science fiction novel? Yu's work speaks a truth about our current relationship with time, memory, and mediation. Today, we all live in a "science fictional" universe, where digital technology has revealed and built vast networks of people, new types of textuality, and whole swaths of distinctive fictions.

This book is a work of television criticism, but it takes into account the important relationship between television content, digital culture, and a quotidian sociability in online media. Today, this relationship comes to a head on our television screens. There has recently been a surge of television programs that feature "temporal displacement"—changes in narrative depictions of time. While forms of temporal displacement have been on television since the start of the medium, there has never been a time with as much temporal displacement on as many shows as there is now. An emergent temporal complexity in online media has engendered this shift in television textuality. Contemporary temporal displacement both amplifies and mirrors the audience's reconstruction of the cultural temporal discontinuity that also emerges on our computer screens. Television thus displaces and distorts the construct of *time,* emulating online media characteristics and serving as a heuristic by which viewers learn to control their own discontinuous lives. By analyzing these *aesthetic* changes, we can uncover the larger *cultural* changes online media create.

Television inherently relies on the concept of time. Specifically, I want to focus on a heterochronic experience of time within television content.[2] As William Uricchio describes, television is heterochronic—displaying multiple time frames—in a number of ways. The first Uricchio calls a "sequence" of television programming; the second an "interpenetration" of "particular programmes over time"; the third the "repetition" of footage or programs. As Sarah Kozloff has shown, time on television is multi-faceted; and ultimately, I agree with Jennifer Gillan, who argues at the opening of *Television and New Media* that television is all "about time."[3] But whereas all three scholars here talk about "time" in the sense of "programming, scheduling, advertising, promotion, and distribution strategies in relation to the evolution of media technologies and viewer practices," I think TV's relationship with time reflects in the textuality of the programming itself. The representation of time within individual television texts relates to larger concerns in the media environment.[4]

Temporal displacement occurs when television programs play with time, using flashforwards, flashbacks, time travel, and/or changes in the protagonist's memory, to heighten the spectacle of the television narrative. Of course, as pop culture writer Steven Johnson has pointed out, narratives always and inevitably "revolve around a mix of the present and future, between what's happening now and the tantalizing question of where it's all headed."[5] I'm looking at long-form serialized narrative, where "the story and discourse do not come to a conclusion during an episode, and the threads are picked up again after an hiatus." [6] Narratives that engage with temporal displacement invoke two or more different time frames, each happening at the same "instant" on screen. They also might engage with characters who have multiple or vast flashbacks (or flashforwards), transporting the plot from the "present" moment to a past or future moment (what Bordwell tellingly describes as "self-conscious narrational asides to the spectator"). [7] Through the deployment of time as a tangible element of a narrative, television producers both emulate and project audience engagement with the television text. I will show that temporal displacement gets the audience more involved in the narrative, allowing them to piece together aspects of the plot. As media content becomes spread out across technologies, as narrative content stretches across episodes, and as representations of characters' memories fracture a program's plot, viewers become assemblers.

Temporal displacement can play with time because it offers viewers greater control of their own linearity. Temporal displacement situates moments of time—past, present, future—within a stable set of bounds. Media themselves lock and are locked: the past is what was on television before; the present is what is on now; the future is "next on." Through the stability of the temporal location of the television *show* on TV each week, the play of temporality within a show itself can become liberating.[8] It can give the audience the semblance of control over time: it seems to say, *if you can understand* Lost's (2004) *bizarre time jumping, then you can control your understanding of any temporal discrepancy.* And in a very real sense, that control represents audience power.

The television show *Flashforward* (2009) illustrates the idea of temporal displacement I discuss in this book. In the show everyone on Earth blacks out simultaneously for 137 seconds. During this period of stupor, the population of the planet "flashes forward" and experiences 137 seconds of their life six months in the future. Upon awakening, each person remembers his/her future, or at least their experience of that future, and then tries to understand it. Some characters are pregnant, some are being held captive, some are drunk, and others are engaged in illegal or illicit activities. The few that didn't "flashforward" believe that their lack of vision signals their imminent death. Yet, as the six-month window closes, characters' realities change noticeably from what their vision indicated. For example, FBI agent Marshall Vogal dies before his flashforward comes "true," illustrating the unpredictability of the future. The development of the narrative towards (the inevitable) moment represented in the characters' flashforwards creates an infinite number of possible narratives, as each character's flashforward becomes a point of reference for a number of "possible worlds"—each narrative thread both existing (in flashforward form) and *not* existing (because it could be changed) at the same time.[9]

The show illustrates temporal displacement by focusing *both* on the narratives of the future and the narratives of the present. Both narratives take place at the same "time" on screen. In each episode, a character's flashforward might be expanded, allowing the audience to see more of the future. The "story" of *Flashforward* takes place between the characters' possible futures and his/her current investigation of that future. Take, for example, the character Olivia Benford, who experiences a flashforward and sees herself talking intimately to an unfamiliar man. When she later glimpses that man in a hospital hallway, she both knows and does not know him—she knows that they will one day know each other, but she does not yet know him. The audience must keep character traits, time frames, and dual narratives in mind simultaneously in order to piece together the show's overarching narrative.

Similarly, Olivia's husband Mark Benford illustrates a second dimension of temporal displacement I discuss in this book: a shift in the characterization of memory. Mark's flashforward takes him to a future in which he is drunk and hallucinatory. Unable to piece together his experiences, Mark's vision of himself drinking undermines his resolution to remain sober. Because his *memory* of a possible future is also affected by that future drunkenness, the images in his flashforward become unreliable and untrustworthy. Much as with Olivia's vision, the audience of *Flashforward* has to sort through aspects of a future that might or might not be "true" (i.e., that actually happen in that future) or can contain aspects that are "false" (i.e., that are manifestations of a drunken vision). Mark must question his own past memory of this future event. Clarifying the reliability of a character's subjective vision necessarily involves intense attention to detail and a powerful understanding of narrative complexity—skills, I argue, that are enabled by today's digital culture.

This concept of "digital culture" manifests through the rapid expansion of online technology. Digital culture inherently relies on electronic communication as a primary means of translating information from one location to another, uses electronic communication to connect with other people from around the world, and creates unique cultural artifacts using digital technology.[10] People from around the world use digital technology to text, to Facebook message, to Tweet, to share on Tumblr, to "check-in"—and often all from the same device, the cell phone. But in all these activities, with all these different technologies, the basic communicative intent—to share with others—doesn't change. In other words, digital culture uses electronic technology to augment, but not to replace, older means of communication.

Yet, we've moved into a new era of connectivity. The default position of our media is "always connected," and we are constantly in contact. Children are growing up in a digital culture using wireless devices, ubiquitous Wi-Fi, and streaming video, and are redefining concepts like privacy, authorship, trust, reputation and textuality. Our conceptions of how to communicate, when to communicate, and what it means to communicate have always been variable. The shift from one communication era to another can be uncomfortable, disquieting even: charges of a decline in morality or a change in the value system of the youth often accompany the shift. We are in the middle of this type of shift, and we can see repercussions shockwave through our culture, from a focus on social media in the news to alterations in our more traditional media forms.

Transgenic Media

These structural and aesthetic changes in television content have appeared at the same time as the emergence of a similar structure and aesthetic in online web media. Terms for online media are varied and ubiquitous. For my purposes in this book, I want to introduce the term "transgenic media" to describe the specific function of contemporary online media as it applies to my argument. The term "transgenic" comes from biology, and refers to the act of inserting new genes into an organism that inherently changes the workings of that organism. And in this case, I use the term deliberately to mean this type of media enacts *new* structures and aesthetics, while remaining tethered to *old* processes and uses. The insertion that changes the workings in the case of transgenic *media* is the activity of a participatory audience.

"Transgenic media" thus refers to the specific type of online/digital/social/new media that has become influential in the past ten years of our culture. Tumblr offers a perfect example of "transgenic media." A microblogging service on which users can upload multimedia from a variety of sources, Tumblr users share television content, filmic content, and other online content in one place, reblogging this content with other users. Tumblr is merely a template, a digital storehouse for information from other technologies; by far, the most common objects reblogged come from older media technologies. For example, a Tumblr titled "Women Reading" consists of users uploading pictures from magazines, from television, from film, and from candids of women sitting in public reading books.[11] The "transgenic" elements here are the displaced *content* from one medium to another combined with the new media *intent* to share this content: the movement of texts from traditional media and put in a new media context with the deliberate intent to share.

I want to differentiate this term from all web content, however—transgenic media is specifically related to online media that invites user participation, as opposed to "classic" web content, which tends to remediate print aesthetics.[12] Besides Tumblr, I will discuss the use of the following technologies: blogs (Blogger, Wordpress, LiveJournal, RSS-feeds), microblogging (Twitter), social networking (Facebook), wikis (Wikipedia, wikias), and online video hosting sites (YouTube). While each of these technologies differs greatly from the others, there is one constant that I see as spanning across each of them: they all exist as a combination of individual contributions coupled with professionally produced templates. The contributions that the users provide become the focus of each of these technologies, whether or not that contribution was made by or is simply being distributed by the user. So, someone posts a short Flip video of their cat ice skating to YouTube, it gets linked to on Facebook, someone Tweets about how funny it is to see a cat on ice, which gets aggregated by a friend's blog, and is then listed on Wikipedia as the first and most classic "cat skating" video in the pantheon of cat skating videos. And is blogged and reblogged on Tumblr. Or, someone records the first ten minutes of *Dancing with the Stars* (2005) and the same process happens. The technologies are about the *distribution* of content, not the *production* of content; the audience becomes the new, transgenic element.

Transgenesis implies both a movement ("Trans" meaning the movement from one place to another) and a novelty ("genesis" being a "rebirth" of sorts), and I use this term not to artificially devise new terminology for online/digital/social/new media, but rather to indicate a further refinement in the way we speak about the everyday media that has pervaded our culture. In the same way we can speak about "cinema," "film," "movies," or "motion pictures" with different connotative meanings, we can speak of "social media," "digital media," "new media," and "transgenic media."[13] For instance, "social media" focus on the way online communication fosters and facilitates human interaction. From the earliest cave paintings, created to tell stories, to the most complex of cinematic narratives, media exist to be shared. Families used to sit around the radio together to listen to broadcasts; the cinema is a shared experience in the theatre; we share links to sneezing pandas on our Facebook page. But the "social" in social media indicates an emphasis on sociality engendered by the technology. Similarly, the term "digital media" describes the *format* of the media rather than its particular use. And "new media," as Lev Manovich has pointed out, is a term that crosses time: every technology is, at some point, "new." [14] Further, every "new" technology is treated similarly within the traditional: fear and resentment at first, followed by (often grudging) respect and then numerous attempts to work with/in these new media.[15] Levinson, for example, argues for new terminology for the spate of interactive websites that emerged *after* the dot-com explosion—"new new media."[16] Ultimately, everything is "new" until something newer comes along.

The point is, the term "transgenic media" is a further refinement, a concentration on the interaction between the new and the old. Is the term "transgenic media" problematic? Of course—no term is going to encompass everything that we need these terms to do, and be entirely unique in its formation. Transgenic media may suffer from the same complexities as I describe above: it may be too limiting or it may be too variable. Yet, for my purposes in this book, all the media I describe enact the "mashup" of producer and viewer that "transgenic" indicates.

Mashup Television

The mashup, a pattern of contemporary media development, shapes television content today. As Gauntlett describes, the interaction between television and online media has shifted significantly, as "the Internet has not only become hugely important in itself, but has forced all the media around it to change accordingly."[17] Today's television programs "mashup" characteristics of online media with characteristics of traditional media. This goes beyond the use of digital on-screen graphics like Twitter hashtag labels overlaid on a program, and instead is more intertextually determined. For example, the NBC comedy *Community* (2009) has integrated multiple mediations in the show, including animations, Claymation, anime-style sequences, homages to alternative media texts, and even the creation of an entirely fictional television show (*Inspector Spacetime*), replete with examples. One could make the claim that *Community* is a televisual example of Tumblr, a mashup of multiple pastiches and parodies of other media. Importantly, such a mashup inherently rests and relies on the audience understanding and constructing the connections between elements.

The mashup takes data from two (or more) different inputs and mixes them together in such a way that a unique, third form outputs without losing the inherent characteristics of the inputs. In my previous work, I have explored the concept of the mashup as a metaphor for contemporary media studies: that is, the mashup is not just a *thing* (an album, a video), but can also represent a *process*, a way of looking at today's media environment.[18] As Spigel claims, the combination of the Internet with the television set has produced a newly engaged audience:

> Changes in programming and sponsorship are met with similar changes in the entire culture of "watching TV." As networks make way for the Net, audiences dip in and out of an immersive sea of interface postings, surfing past the interplay of news headlines, TV promos, chat rooms, media criticism, weather reports, advertising pop-ups, and self-help columns as these flash across the increasingly segmented computer screen.[19]

Through intense interactivity, deep engagement, and a solid immersive foundation, television narratives that engage with temporal displacement do so not simply because of online media, but also with the same mechanisms and focus. By provoking, and indeed, illustrating, a profound *need* for time, contemporary television ushers in a new perspective on cultural temporality.

Most commonly, mashups can be seen across the diverse fields and disciplines of popular music, software programming, and online videos.[20] Contemporary artists have been using samples of other artists' work as reference within their own music for years (like sampling in Hip-Hop music). Mashup songs use a complete remix of music from other artists to create a new sound. For example, Girl Talk mixes hundreds of songs to make a fresh sound that both references but also subverts the meaning of the original work. Another classic audio mashup is DJ Danger Mouse's *The Grey Album*, which mixes The Beatles' *White Album* with Jay-Z's *The Black Album* to create a unique, third voice. We listen to *The Grey Album* and hear *both* the original Beatles and Jay-Z music *and* the amalgam of them as a new form: a holy trinity of meaning.

In software studies, alternately, the term "mashup" can refer to software that uses or remixes data from two or more online sources to create a new service: for example, Earth Album mixes the software for Google Maps with the software from Flickr and YouTube, so that users can logon to Google Maps and view images and videos that have been "tagged" from these locations.[21] Amber Davisson argues that Google map mashups played an important role in the 2008 Presidential election, for instance, leading to a larger cultural innovation in our understanding of the role of civic engagement.[22]

Video mashups have become integral to the contemporary media environment, and have become fixtures in discussion about social media.[23] Indeed, Chuck Tryon illustrates the crucial role that mashups have played not only in the legal fields of copyright, authorship, and fair use rules, but also in political campaigns. Mashup videos that remix political content function as "satire, commenting on current events rather than remaining content to focus on genre conventions."[24] These political mashups often take audio material from one source and mash it up with visual content from another, as in David Morgasen's 2008 video, "Obama and McCain- Dance Off!"[25] In this video, impersonations of Barack Obama and John McCain are mashed-up with images of the politicians' faces superimposed on dancer's bodies. The 2008 election between Obama and McCain becomes fodder for a "dance-off," in which both candidates compete to see who has the "sickest [dance] moves." Morgasen effectively satirizes the spectacle of the 2008 Presidential election through this juxtaposition, calling into question our delineation of entertainment and news.

Contemporary temporal displacement on television enacts another type of mashup. Using simultaneously occurring, dual narrative time frames, contemporary television shows play a crucial role in the popular understanding of audience interaction with online media. In other words, by simultaneously seeing the past and the present, the past and the future, or the present and the future depicted on screen, television "mashes up" two time frames to create one complex narrative. The "new modes of ... television-Net convergence" that John Caldwell describes thus have not only had "an impact on television's textual forms and the ways we relate to television itself," but have also impacted larger cultural issues as well.[26] The audience's construction of television narratives from these dual time frame/fragmented plots serves as a metaphor for the way we construct coherence out of our own fragmented lives.

An example of the mashup-as-temporal-displacement appears in the *Boston Legal* (2004) episode "Son of the Defender," which mashes up footage of actor William Shatner from 1957 with footage from 2007. In this episode, *Boston Legal* reused existing footage from "The Defender," a 1957 episode of *Studio One in Hollywood* (1948) as a flashback for protagonist Denny Crane (William Shatner).[27] Shatner portrayed the protagonists of both shows, and "Son of the Defender" makes the argument that the character from "The Defender" is actually a younger version of the character from *Boston Legal*. By integrating both television texts into the same narrative, *Boston Legal* not only mashes up two different television programs, but also temporally displaces the character of Denny Crane by inherently relying on the audience's knowledge of television as a historical document. Obviously, audiences don't have to know that the original episode was named "The Defender," but the reference to the *Boston Legal* title "Son of the Defender" carries more weight when audiences make the realization. Additionally, Shatner's own celebrity becomes influential here. Both characters become subsumed under Shatner's contemporary star persona, as well as the audience's knowledge of the younger "star-to-be." By combining two texts into one, the character Denny Crane becomes integrated into the "Defender" character Kenneth Preston, and yet each must remain divisible from the other in order to grasp the larger meaning of the combination.

Thus, I show that the concept of the "mashup" represents a new metaphor for the relationship between television content, audience expectations, and character within narrative. We can arrive at a deeper meaning for the spectacle of temporal displacement, both on television and in our own lives, through the recognition of the individual "inputs" that mashup to form these new outputs.

Conclusion

Each section of this book ties together a diverse set of contemporary television programs that illustrate temporal complexity in a variety of contexts. I examine individual television texts to explore larger issues in media studies. The assertions I make about television media elucidate cultural issues that are also reflected in our contemporary usage of transgenic media. Each chapter of this book examines one aspect of temporal displacement by analyzing the symptomatic manifestation of those issues on temporally displaced television. The next chapter in this opening section details more specifically the relationship between television and transgenic media. In the second section of this book, "Memories," I examine the way transgenic media both reflect and build upon our individual memories as demonstrated by characters within television shows. In Chapter Two I introduce the concept of "mediated memory" as a central component of contemporary television. I argue that much of our contemporary individual memory is connected to the media, and that our media also reflect the concept of memory back at us. Similarly, today's online media both *externalize* our memory by recording and displaying a complete history of the self and *internalize* our memory by offering a technological means for reinterpreting the self. I examine television that plays with the idea of memory as an extension of a lived, temporal existence.

In the third chapter, I specifically examine the BBC television shows *Life on Mars* and *Ashes to Ashes* as complements to and critiques of a form of mediated memory. In essence, these shows illustrate the power of a personal narrative format. This personal narrative has become a powerful voice for transgenic media users—Wordpress publishes autobiographies, YouTube allows public performance of personal history, and Facebook publicizes our community and culture. Yet, personal narratives also present a troubling glimpse into a mutable history, or an alterable past. Who are we now if we can change the representation of our past self? This duality of transgenic media—its public nature and its private intensity—is revealed through the temporal adventures of protagonists Tyler and Drake in *Life on Mars* and *Ashes to Ashes*.

The next section of the book, "Temporalities," focuses specifically on the narrative implications of time travel as temporal displacement, and the way these implications mirror similar issues in our online media use. As in the previous chapters, television becomes a way to work through issues of temporal complexity, and in this section I examine in detail one duality that seems to be central to our quotidian media usage: permanence and ephemerality. In the fourth chapter, I argue that a resurgence in the time travel narrative on television reflects a renewed cultural focus on the persistence of history within online media. We now have the ability to scroll through our entire digital lives, our history recorded seemingly forever. Through readings of *Doctor Who, Eureka, Flashforward,* and *Lost,* I show how the time travel narrative illustrates this cultural obsession. In Chapter Five, I examine the converse of the permanence of history: how time travel in the recent rebirth of *Doctor Who* relies on the rewritability of history and memory, a notion that is mirrored in our culture's reliance on the wiki as an encyclopedic database. By looking at four different functions of history in the narrative of *Doctor Who,* and then examining how these functions become represented on tardis.wikia (a fan-created wiki about *Doctor Who*), I illustrate a historical editability, and show how traditional media forms can often offer a useful heuristic for understanding online media technologies.

In Chapter Six, I synthesize Chapters Four and Five by examining the metaphor of the network within *Lost* and *How I Met Your Mother*. I argue that both shows represent a form of networked storytelling, resulting from a contemporary focus on the online social network as a space for viewer identification to occur. By examining the role that the network has had, both in our online media use and in our digital culture at large, I illustrate new ways of organizing our impermanent yet recorded personal history.

Finally, at the conclusion of the book, I postulate a new cultural moment, beyond postmodernism, which calls for, among other things, a redefinition of temporal structure. This new era reinscribes meaning to temporal displacement and argues an underlying structure to historicity. The emergence of historical complexity lies at the heart of our fragmented culture. Specifically, by piecing together fragments of a temporally separated narrative, viewers engage with the same types of actions that are also needed to piece together fragments of a spatially separated (or transmediated) narrative. In doing so, audiences mirror a similar "piecing together" of a fragmented culture, which emphasizes a continuity between past, present, and future.

Notes

1 Charles Yu, *How to Live Safely in a Science Fictional Universe* (New York: Pantheon Books, 2010).

2 William Uricchio, "TV as Time Machine: Television's Changing Heterochronic Regimes and the Production of History," in *Relocating Television: Television in the Digital Context*, ed. Jostein Gripsrud (London: Routledge, 2010), 30.

3 Sarah Kozloff, "Narrative Theory and Television," in *Channels of Discourse, Reassembled: Television and Contemporary Criticism*, 2d ed., ed. Robert C. Allen (Chapel Hill, NC: The University of North Carolina Press, 1992), 90–1.

4 Jennifer Gillan, *Television and New Media: Must Click TV* (New York: Routledge, 2010), 1.

5 Steven Johnson, *Everything Bad Is Good for You* (New York: Riverhead Books, 2005), 55.

6 Kozloff, 91.

7 David Bordwell, *Narration in the Fiction Film* (Madison, WI: University of Wisconsin Press, 1985), 79.

8 Even if timeshifted via online environment or DVR, the show is still experienced in a specific order, in a specific format, and in a specific structure.

9 Marie-Laure Ryan, *Narrative as Virtual Reality: Immersion and Interactivity in Literature and Electronic Media* (Baltimore, MD: The Johns Hopkins University Press, 2001), discusses the notion in fiction that any opportunity left unexplored opens up a "possible world."

10 See Clay Shirky, *Here Comes Everybody: The Power of Organizing without Organizations* (New York: Penguin, 2008).

11 See http://womenreading.tumblr.com/.

12 See Jay David Bolter and Richard Grusin, *Remediation: Understanding New Media* (Cambridge, MA: The MIT Press, 1999).

13 Paul Levinson, *New New Media* (New York: Allyn & Bacon, 2009).

14 Lev Manovich, *The Language of New Media* (Cambridge, MA: The MIT Press, 2001).

15 Tom Gunning, "Re-Newing Old Technologies: Astonishment, Second Nature, and the Uncanny in Technology from the Previous Turn-of-the-Century," in *Rethinking Media Change: The Aesthetics of Transition*, ed. David Thorburn and Henry Jenkins (Cambridge, MA: The MIT Press, 2004), 40.

16 Levinson.

17 David Gauntlett, "Media Studies 2.0: A Response," *Interactions: Studies in Communication and Culture* 1, no. 1 (2009): 148.

18 Paul Booth, *Digital Fandom: New Media Studies* (New York: Peter Lang, 2010).

19 Lynn Spigel, "Introduction," *Television after TV: Essays on a Medium in Transition*, ed. Lynn Spigel and Jan Olsson (Durham, NC: Duke University Press, 2004), 5.

20 Chuck Tryon, *Reinventing Cinema: Movies in the Age of Media Convergence* (New Brunswick, NJ: Rutgers University Press, 2009), 82.

21 See Martin Brinkmann, "Six Great Google Maps Mashups," ghacks.com, 08 Nov 2007, http://www.ghacks.net/2007/11/08/six-great-google-maps-mashups/ (accessed 23 Dec 2010).

22 Amber Davisson, "Beyond the Borders of Red and Blue States: Google Maps as a Site of Rhetorical Experimentation in the 2008 Presidential Election," *Journal of Rhetoric & Public Affairs* 11, no. 4 (2011).

23 Tryon, 158.

24 Ibid., 167.

[25] David Morgasen, "Obama and McCain—Dance Off!" *YouTube*, 23 Oct 2008, http://www.youtube.com/watch?v=wzyT9-9lUyEzz (accessed 01 Mar 2011).

[26] John Caldwell, "Convergence Television: Aggregating Form and Repurposing Content in the Culture of Conglomeration," in *Television after TV*, ed. Lynn Spigel and Jan Olsson (Durham, NC: Duke University Press, 2004), 43.

[27] See http://en.wikipedia.org/wiki/Boston_Legal#Breaking_the_fourth_wall (accessed 05 Apr 2011).

One

Television and Transgenic Media

I am, in a way, telling myself what to think, that my future self has produced a record of the out-put of my consciousness, of my internal monologue. Or rather, a dialogue, between myself and my future self, in which my future self is telling my present self what I have already finished thinking but have not yet realized I thought.

—Yu, How to Live Safely in a Science Fictional Universe, 109fn

Charles Yu's confusingly distorted communication mirrors a similar temporal confusion within the narratives of contemporary television programming. Existing as he does in past, present and future, Yu's journeys across time and space culminate with a message his future self sends back in time to his past self, which Yu reads in the (reader's) present. The message that Future Yu sends, in a meta-moment, is the book *How to Live Safely in a Science Fictional Universe.* Yu is then thrust into a temporal swirl of past, present and future, each of which converges on one particular moment in time—the precise moment the reader is experiencing as well. The novel's temporal mechanics focus entirely upon the reader's present, engaging the reader with his/her own vicarious time machines. Television in the first decade of the twenty-first century offers a similarly disjointed framework which centers on the viewer. The complexity of today's temporally displaced television narratives illustrates multiple subjective viewpoints. The enigmatic narrative has become common, as viewers actively work to construct narratives, a characteristic that particularly appeals to cult television viewers.[1]

Depending on whether we're discussing *content* or *technology*, the term "television" can be highly fluid. "Television" has most commonly referred to the box that sits in many of our living rooms, kitchens, dens, bedrooms, and (occasionally) bathrooms. When I use the term "television," however, I'm specifically referring to the structural and aesthetic *content* that (until recently) one would watch on that box in the living room. Even though we can watch on our iPods, our Droids, our iPads or our netbooks; even though broadband speeds have increased and computers have become more prevalent; even though much television viewing has shifted to online venues like Hulu, YouTube, and Netflix or even smart phones or tablet PCs—in this book I am tautologically considering *television content that has been created to be on television technology* (that is, I'm interested in the content regardless of whether or not it *can* be online). And I'm interested in the multiple paratexts, ancillary texts, secondary texts, and mashup texts that become part of the transmediated television text.

I want to make the connection between the form of temporal displacement witnessed in contemporary television and a similar postmodern fragmentation perceived in our own lives, as we eke our way through a highly mediated twenty-first century. The temporal displacement of television's complex narratives does not happen simply by what we witness on the television screen, but can also occur through the complex interactions between numerous other devices and technologies that continue these narratives across media boundaries. In contemporary television studies, much has been written about this concept of "transmediation," the movement of narrative meaning across media boundaries.[2] We assemble our transmediated narratives together as if they were constructed like a jigsaw puzzle.[3] Transmediation can subvert the traditional linear order of beginning/middle/end, or start to finish, for we may encounter bits of this puzzle out of order: for instance, watching the webisodes of *Battlestar Galactica* (2004) after we watch other episodes.[4] Or, we could piece together a transmediated franchise like *Heroes* (2006) via multiple media outlets, like the television program, graphic novels, mobisodes or webisodes. Our temporal relation with the narrative of *Battlestar Galactica* or *Heroes*, therefore, is only partially based on the discourse of the show—it is dependent too on the interactive mechanics of the transmediated text. Although all narratives are constructed from a variety of elements, a transmediated narrative necessarily breaks technological and temporal boundaries. Transmediation, therefore, represents one way of examining contemporary narrative structure that takes into account viewers' own timeline (their temporal perspective).

But it's important to recognize that not all media content is transmediated—in fact, a majority of contemporary television is firmly fixed into the televisual medium. The *aesthetics* and *structure* of transmediation, however, appear across a range of programming. While transmediation is the merging of *content* across *technologies*, many serialized television programs merge their *own* content and themes within the same technology, using the same "jigsaw puzzle" mentality as a transmediated text does. The "point" of many temporally complex television shows isn't just following the narrative, but actually piecing it together entirely. The contemporary media environment highlights transmediation not just as a style of mediation, but also as a metaphor for larger cultural shifts in the media environment.

This isn't to say that the mashup of online technology use has haltered or changed the larger hegemonic influence of political or social mores on the media output of our culture.[5] The "interactive" use of online technology is itself delimited by the types of traditional media viewed on television. There are few "experimental" or "avant-garde" videos online; any user-generated content tends to parallel the culturally acceptable (and popular) content seen on television, film, and other media. Indeed, this is perfectly illustrated by "The Trailer for Every Academy Award Winning Film Ever," which uses online video to parody Hollywood convention.[6] The parody was uploaded by users (not paid producers) of video, but it works in a similar way to other traditional media, even as it parodies them.

My argument, then, is merely that the mashup of online technology with traditional technology reflects a growing change in the way our culture understands and deals with time, memory, and history. The increased use of online technology has intimately changed the audience's relationship with time in their own lives. In effect, we "transmediate" every time we go online or participate in digital communities. In 2010, as Kristen Daly argues:

> interaction becomes a natural consequence of digital media. Computer and digital users have been trained by their immersion in digital culture to participate in/with what they consume. With the disappearance of the spectator, what becomes of the spectacle? Instead of world as picture we have world as game.[7]

This focus on the ludicity of contemporary media lies at the heart of my previous book, *Digital Fandom*.[8] In it, I established that social media rework the temporal arrangements of other media in both reception and production arenas. Locations where reception is intimately tied to production, like interactive websites such as Wikipedia and RSS-powered news or blog feeds like Google Reader, are inherently unstable entities: what is posted one moment can be changed the next. For example, news-aggregating sites value stories based on popularity: a popular story such as "Understanding The Dangers of Radiation Exposure" can be replaced moments later by "The 7 Most Heroic Con Artists of All Time." With seemingly little individual influence over the stability of the medium, users of interactive websites sacrifice temporal control for immediately updatable texts.

In *Digital Fandom* I also examined how the advent of social media in general, and online fan communities in particular, necessitated a change in the way we study media. In this book, I want to augment that argument by examining the way transgenic media affect the "traditional" medium of television. If *Digital Fandom* was about audiences, here I'm talking about texts. In many current examinations of the interaction between digital technology and television, research has tended to focus on the way digital technology manifests characteristics of traditional media.[9] Perhaps the most salient and well-honed example is Jay David Bolter and Richard Grusin's *Remediation,* in which they describe how all media is remediated from older media content.[10] Furthermore, in *Television after TV,* Caldwell discusses the way websites work with and as ancillary products for television programming. He specifically examines *Dawson's Creek* (1998) and looks at how the web gives access to the "personal lives" of the characters through "'characterized' proliferations of the text, 'narrativized' elaborations of the text; 'backstory' textuality; 'metacritical' textuality; technological augmentations; and merchandizing augmentations."[11] Additionally, in Anne Friedberg's contribution to *Rethinking Media Change,* the influence of *television* on *digital media* is made clear, as she compares the computer screen to the television screen as "a component piece of architecture, a 'virtual window'...of built space."[12]

However, little research exists that examines how digital technology can affect the *content* and *form* of traditional media. The research that does exist usually views specific instances of influence, not larger cultural issues. For example, Boddy describes how the introduction of the Personal Video Recorder (PVR) helped usher in a new industrial model to the content of television.[13] Further, in *Show Sold Separately,* Jonathan Gray elaborates on how television gains additional meaning from the paratextual web of content that revolves around it.[14] "Television" has even come to mean aspects of content that aren't even typically viewed on television—advertising for programming and technology, print ads about television shows, and even tangible goods like action figures and calendars. These paratextual elements help form our opinion about the concept of television before we have even turned the set/box/computer on.[15] Gillan charts the connections between individual media texts and their social media outlets—for example, 24 (2001) and its mobisodes—

and the way the television industry uses social media to describe alternate viewing patterns.[16] To understand this moment of media confluence, it's important to recognize what Thorburn and Jenkins argue about media in transition: "each medium is touched by and in turn touches its neighbors and rivals."[17] Digital technology touches television, affecting not just the industrial development of the medium, but also its structure and aesthetics.

Postmodern "Malaise"

The cultural theory of postmodernism is a useful, if rather broad, starting point for analyzing this medium interaction. Postmodernism can be seen as a highly mediated theoretical structure and reminders of the mediated nature of postmodernism are all around us: for example, in "Operation Righteous Cowboy Lightning," a 2011 episode of *30 Rock* (2006), Liz and Tracy argue while Tracy is being filmed for a reality television show. Liz and Tracy refuse to forgive each other, hampering the attempts of the reality show producer to tack a happy ending onto the episode. In order to provide their audience with closure, the reality TV producers record two Liz and Tracy lookalikes embracing. The real Liz and Tracy watch this fake ending, and are so moved that they are able to forgive each other in "real life." But instead of *truly* making up, they simply shrug and remark that there's no need—now that they already saw their reconciliation on television. As an example of Baudrillard's hyperreality, this episode of *30 Rock* almost perfectly encapsulates the notion that what is "real" is just an illusion of reality.

Temporally complex shows focus on temporality in part as both resistance and panacea to what postmodern theorists like Fredric Jameson refer to as a cultural "schizophrenia."[18] In turn, the skills viewers use to create a whole from a fractured narrative help structure his metaphor of schizophrenia. For theorists of the postmodern, the contemporary era creates a certain temporal misalignment. There is no sense of the "past" or "future," but rather an instantaneous and vacuous sense of the "present" (as Yu described in the inscriptions to this and the previous chapter), as postmodern schizophrenia is characterized by "the absolute proximity, the total instantaneity of things."[19] Postmodernism sees our contemporary era as an age of über-simultaneity where we experience many different temporalities all at once as a "temporal malaise."[20]

By examining the past decade's eruption of temporally complex television shows across a range of genres—shows such as *Ashes to Ashes* (2008), *Doctor Who* (1963–1996, 2005–), *Dollhouse* (2009), *Eureka* (2006), *The Event* (2010), *Family Guy* (1999), *Flashforward*, *Fringe* (2008), *Heroes*, *How I Met Your Mother* (2005), *Life on Mars* (2006), *Lie to Me* (2009), *Lost*, the *Star Trek* (1966-1969, 1987–1994, 1993–1999, 1995–2001, 2001–2005) franchise (among others), I want to argue that the "schizophrenia" these theorists describe becomes a metaphor for how audiences respond to the temporal complexity in television narrative, and within digital culture. The temporal complexity of the contemporary television show leads to both an aesthetic appreciation of the narrative and a feeling of disorientation within an already fractured postmodern life. Audiences learn to reconstruct temporally fractured narratives and apply that sense of reconstitution to their own postmodern lives. Of course, postmodernism (by definition) cannot be a totalizing force for understanding what is at work here: I don't mean to overplay the postmodern card, and the majority of this book will examine specific texts rather than overarching theory. Understanding how viewers engage with temporal displacement in their quotidian lives is a useful heuristic for examining audiences' temporal instability and the contemporary media industry's turn towards temporal displacement. I examine this temporal complexity in all its guises through a rhetorical/cultural point of view. Using the detailed study and close readings of texts, cultural practices, and previous theory to analyze the content of television can contribute to and enrich a larger understanding of the role of television in a variety of settings.

The Aesthetics of Contemporary TV

Not all shows demonstrate temporal complexity (even complex shows like HBO's *The Wire* (2002) are temporally quite basic), and I do not intend to suggest that temporal displacement is unique to the past ten years of television.[21] But temporal displacement leads to a fascinating new turn in the aesthetics of television. Indeed, classic sitcom *I Love Lucy* (1951) used flashbacks, and *Newhart* (1982) famously ended with a surprise "this-is-all-a-dream" ending that flashed-back to Bob Newhart's earlier *The Bob Newhart Show* (1972). These shows, however, merely featured temporal displacement in an episode or two as unique moments. Sconce describes these unique, one-off episodes as common in television, to the point that we recognize the common "amnesia" plotline as a reoccurring trope across texts. What Sconce describes are momentary narrative shifts in what are otherwise recognizably consistent programs. That is, audiences accept a one-off "it was all a dream" episode because they know that the show will return to "normal" next week. For example, in the episode "Bombshells" of *House* (2004), protagonist Cuddy is admitted to the hospital, and while anesthetized, she hallucinates a number of "what if" moments that, as Sconce describes, allow the audience to imagine new possibilities in the text—House and Cuddy as Butch and the Kid, or House and Cuddy in a cabaret musical number.[22] The audience can imagine these new possibilities because the show returns to "normal" at the end of the episode, when the hallucinations end.

But in the past decade the number of shows that seem to *rely* on these differentiations as an aesthetic mechanism has greatly increased, signaling a cultural acceptance of the aesthetics of transgenic media. For example, *Dollhouse* viewers explored the weekly erasure of the protagonist's memories and personality—*every* episode was an "amnesia" plot. The main characters were eternally displaced, and until their personalities had stabilized, each narrative reintroduced each character.

As Michele White argues, today's multi-media environment links television studies to new media studies in both aesthetics and content.[23] It's not that viewers, users, or even "viewsers" are increasingly becoming less capable audiences, but rather that our emerging digital society is subtly changing viewing habits and between Jack's experiences on the island, and his life off the island at "Hoffs-Drawlar Funeral Parlor." The oddly named location offered an anagrammatic clue to a change in the temporal shifting: Hoffs-Drawlar rearranges to spell "Flash Forward."[68] Indeed, the flash*backs* that many in the audience thought they were watching for the third season were revealed in that last episode to be flash*forwards*, taking place in the narrative future, a temporal displacement that could only be answered with an operational aesthetic "wow."

Conclusion

In this book I will explore what it means to live in a temporally complex world, as seen through our television narratives, our online media, and our digital lives. I want to look at how television producers "play" with the textual existence of temporality in television narratives to create *affect* in the viewer: to give the viewer this "wow" moment of narrative.[69] This book is an attempt to expand scholarship on aesthetics and narrative form on television, an area of television scholarship with vast uncharted regions.

Why is this so important? If the past decade has seen the emergence of an incredible array of new technologies for mediated entertainment—iPads, DROIDs, Netflix Streaming, Xbox Live, Facebook, Twitter, etc.—then the choices of where audiences receive their entertainment are multiplying at an almost exponential rate. Regardless of technological device, however, the *content* is still remarkably similar across technologies. According to a 02 Jan 2011 *New York Times* article, television is still the most viewed medium in the household, with the average American watching over 34 hours a week.[70] Indeed, a 2009 study by the Nielson Media Research revealed that television content (whether viewed on a traditional television screen or on a computer or mobile device) is the most viewed type of content in the US.[71] As technology diversifies, so do the focus and the possibilities of television content. Television used to be ephemeral: before effective recording technology like VCRs, viewers could only watch when a show was broadcast. Miss an episode and, unless it was repeated (which it would rarely be), one was usually out of luck audience's expectations of media.[24] Although many factors influence this increase in contemporary temporal complexity—including the advent of new technology, new industry practices that take account of these technologies, and global media conglomerates that find financial incentive to pursue them[25]—I want to focus on the aesthetic and structural changes engendered by online social technologies. For David Thorburn, investigating aesthetics is not simply a value-laden venture, but rather one that is central to exploring the televisual medium.[26] As he states, "we must be able to read [television] texts in something of the way the audience experiences them: as stories or dramas, as aesthetic artifacts, whose meaning will be fully available only if we employ, along with other interpretive methods...strategies of reading."[27]

Importantly, as Sconce argues, television studies tend to elide analyses of aesthetics in favor of examinations of the television industry.[28] Yet, television aesthetics do have an important role to play in investigations of the contemporary media environments, and "we must be explicit in acknowledging the roles of evaluation and aesthetic judgment" in our scholarship.[29] What interests me today is what Jason Mittell describes as the particularly unique demands of narratively complex television: "that viewers pay attention more closely than typical for the medium" and that "the poetic techniques that programs use ... engage viewers and enable long-term comprehension."[30]

Other contemporary books on television programming, like Shimpach's *Television in Transition,* focus on different aspects of contemporary television content, including a "seismic shift in the institutional practices of a television industry increasingly concerned not only with the small screen but also with the whole world and multiple media outlets."[31] As he makes clear, this form of television cannot be separated from the industrial infrastructure.[32] But to research one is to elucidate the other, and any industrial analysis can be complemented by textual analysis.

Shimpach's analysis is mindful of the concept of "flow," Raymond William's 1974 exploration of television aesthetics.[33] Flow crucially demonstrated how the *style* of television was unique to the medium. Defined as the interplay between programming, publicity, advertisement, and service announcement, the flow of one element into another helped to create a sense of normalcy in the media. Yet, it can still be difficult to pin down flow's application in new contexts.[34] Indeed, as Sconce points out, "even for scholars conversant in televisual textuality, this emphasis on flow can, somewhat ironically, render the medium into the very same hallucinatory chaos experienced by Williams himself that night in the Miami hotel room where he first conceptualized the term."[35]

That being said, flow has proven useful for many scholars—for example in relating gender and television[36]—and was crucial for understanding the way people watched television during Williams' writing period. Of course, given the rapid diffusion of VCRs, DVRs and DVDs, the concept of flow has radically shifted from Williams' influential study, as:

> this new form of time-shifting is merely one sign of the ways in which digital technology, at least in the eyes of many current industry leaders and pundits, is eroding the experience of simultaneity and liveness that has traditionally been seen both as part of television's essential nature and central to its relation to the nation.[37]

As William Boddy further points out, "at least some of TV's new digital technologies are already having an effect on program forms and advertising practices," one of which is the concept of the flow of information.[38] Amanda Lotz points out how flow needs a "significant revision" in the 21st century, as "the continuous infiltration of control devices into television use has greatly disrupted flow as a fundamental characteristic of the medium—at least in terms of television flow being determined by someone other than the individual viewer."[39]

Yet, what Boddy and Lotz both assume, reading quite accurately from Williams' original assertion, is that changes in flow are dependent on television *technology*: that is, the development of the television as a device of mediating content. The remote control, DVR, DVD, and other technological devices like mobile apps and on-demand entertainment, have certainly affected the technological development of the television, which has in turn changed the concept of flow. Gillan even notes that in the 21st century, flow involves moving *people* from the television to the Internet.[40] But beyond technology, I believe that the actual *content* of television itself has shifted in the past few decades, changing the concept of flow for the 21st century.

Specific discussions of the aesthetics of television have been more circumspect about delineating particular stylistic concerns, not surprisingly because of how flow has dated in the years after its discussion. For example, Caldwell establishes a new sense of television style based less on a static sense of classical aesthetic tradition, and more as an "activity—as a performance of style."[41] For Caldwell, television style, what he calls "televisuality," is based on a consistently *shifting* sense of style. That is, television style is always in flux. More recent works have focused on the relationship between the web and television. Michael Kackman, et al.'s *Flow TV* argues that in the 21st century, issues of "flow" have shifted to issues of "convergence," where the interplay between different media technologies takes precedence.[42] Sharon Ross' *Beyond the Box* takes a look at the way individuals participate with television texts online, examining the "tele-participation" of fans with (mainly cult) television texts.[43] Much of her discussion centers on the way fans are encouraged/manipulated into interactions with the television text through Internet activities, what she calls "invitations."[44] Glen Creeber, further, notes how contemporary long-form drama can add a sense of "realism" to television by referencing real-world networking, a concept I explore in Chapter Six.[45] All television, according to Uricchio, is a "constant remix of representational and temporal elements," presenting real-world material in new contexts.[46]

The rapidly changing face of television is nowhere more evident than in literature that focuses on the intersection of television and the web. Internet technology like streaming video (Netflix, Hulu, or

Vudu) has made the connection between the web and the television more closely aligned. Janet Murray argued in 1997 that online media would form a "hyperserial," which "would offer a complete digital library of [a television] series, and these episodes...would be searchable by content."[47] The existence of Netflix Streaming and other such services validate her claim. Ellen Seiter argued in 1999 that "television sets and computer terminals will certainly merge, cohabit, and co-exist in the next century."[48] While her vision may not have completely emerged during the first decade of the 21st century, it is well on its way. The structural changes to television create an "increasingly multidimensional and fractured" image.[49] As Jonathan Hardy indicates, however, there has been a turn in contemporary television production, from a specific concentration on television content to one that looks at a universe of intertextual content.[50] Gray reframes the television program as outside the focus of television studies, and argues that we can study television through its satellite components: even our very understanding of what *television* is becomes a "mashup" of different elements.[51]

Temporal Displacement in Practice

The third season finale of *Fringe* puts into practice the two types of temporal displacement I describe in this book: memory temporality and character temporality. For most of the third season, the "Fringe" team of scientists and agents (an elite team of paranormal researchers) has been investigating a universe parallel to ours. One character, Peter Bishop, assembles a machine that can traverse the two. The machine acts as a link between our universe (Universe A) and the alternate universe (Universe B). When he first steps into the machine, Peter's consciousness travels fifteen years into the future and he discovers a dystopian, apocalyptic world in which Universe B had been destroyed and in its wake Universe A was collapsing. The first type of temporal displacement—memory temporality—becomes apparent here when Peter's mind (although not his physical body) is transported to the different time frame and back and the audience must piece together events from the future (that haven't yet happened) with those from the present (as they are happening) as he remembers them. A second type of temporality—character temporality—also becomes clear in this episode of *Fringe* when Peter's father, Walter, learns that the only reason Peter is able to assemble the universe-traversing machine is because a Future Walter (one who has not yet existed) sent one back in time for Peter to find. As Walter says:

> I knew the pieces were buried millions of years ago, but how did they get there, so deep in the past? But now I understand. I sent them there. The wormhole in Central Park—I sent them back through time—I have already done it. Therefore, I have no choice but to do it again.[52]

The shift in verb tense in his dialogue indicates the temporal displacement: even though Walter (from his perspective) hasn't sent anything through time, because they *have* been sent through time he knows they *must* be sent through time. And if he doesn't, there will be a paradox, a logical inconsistency in the plot. The mental time travel that Peter endures is matched symbolically with a physical time-travel that Future Walter must create with the machine parts.

Although "The Day We Died" is an episode of *Fringe* that ties both types of temporal displacement together, most often on television they are seen in different ways. Two recent television programs, *The Event* (2010) and *The Chicago Code* (2011), contrast these forms of temporal displacement. *The Event* uses a temporally displaced narrative to detail the events of an alien invasion. The opening few episodes are depicted in a temporally complex manner, with scenes appearing out of chronological order, effects happening before causes, and a narrative style occurring disjointedly. In contrast, *The Chicago Code* shifts the mode of memory temporality through a novel use of flashbacks that highlights the role memories play in telling the nonlinear story of police corruption in Chicago. Although the story is rarely told out of chronological order, and the events happen as one would expect—causes before effects, crimes before punishments—*The Chicago Code* does offer temporal discontinuity through characters as they relate to the audience. In short, like *Fringe* both shows tell a complex story. But whereas *Fringe* uses actual time travel (both in the mental and in the physical sense), both *The Event* and *The Chicago Code* mash up narrative and character to play with time narratively, and to engender audience identification.

In the pilot episode of *The Event,* the first image the audience sees is a rapidly expanding digital clock face, symbolizing the importance of time to this narrative. The episode then opens with the climax of the episode's narrative, as a hijacked plane is about to crash into a party in Miami. The scene is depicted as if it was being shot by news media cameras, the images blurred and shaky, pixilated and full of static. The scene abruptly shifts to an airport and a title appears: 23 minutes earlier. Already, in the first 30 seconds of the episode, the temporal structure appears to run backwards. The effect has come before the cause.[53] The next scene of the pilot episode shifts gears and we see the house of one of the protagonists, Leila, and another new time frame, 11 days earlier. The episode continues in this fashion, jumping from time frame to time frame and revealing piecemeal the elements of the narrative until the final climax of the episode, when the story returns to the originating time and location to witness the airplane disappear from the sky. At last, again, could this be the titular Event?[54]

Throughout this temporally discontinuous plot, the audience must continually construct the story from the events as they unfold. At first glance, this is no different than any other show, during which the audience would also have to figure out a plot from separate elements. But because the events in *The Event* are told out of order, there is an extra layer of audience interaction with the text. Not only do we have to construct the meaning from the spatial relationship between characters and places, but we also have to understand the *temporal* relationships, and place characters within a specific time line.

In contrast, the pilot episode of *The Chicago Code* illustrates a sense of temporal displacement in terms of the characters. In this first episode, one of the main characters, Antonio, serves as the driver and confidant to Police Superintendent Colvin. In a subjective voice-over, Antonio tells how he grew up in Chicago, was recruited to the force by the superintendent, and happily serves as her driver. As he speaks about his past, the events are visually represented as flashbacks on screen: child Antonio in his house, young Antonio in a bar, and teenage Antonio on the force. His voice-over of the past, however, is punctuated and ultimately silenced by three gunshots that strike him dead in the present. The flashback he narrates abruptly ends, cutting "back" to the ongoing narrative. Both his voice-over and his flashback disrupt the flow of the narrative and shift from a future-tense narration to a present-tense death. The temporal placement of the voice-over shifts, making it seem as though Antonio was speaking *in the present,* at that precise moment, and confusing the narrative depiction of events. Yet, the moment when his non-diegetic voice is stopped by gunshots also coheres the audience to the character. In that instant, the audience is *transported* to that crime scene because the audience has been following the flashback of that character. We have followed Antonio as he grew up in Chicago, and by marking his death by the absence of his voice, the show makes clear our identification. Both *The Chicago Code* and *The Event* offer a sense of temporal displacement, but each manifest it differently.

Temporal and Narrative Complexity

Such temporal complexity has become common across television shows, enacting three levels of textual "play" with time. First, characters "play" with their memories, evoking an ever-changing personal history that influences the narrative development. For example, in *How I Met Your Mother,* protagonist Ted often mis-remembers aspects of his past, the flashback of which illustrates this forgetfulness. A minor character who appears in a flashback in "How I Met Everyone Else" is named "Blah-blah" because Ted can't remember her real name. Second, temporally complex television "plays" with narrative, forcing the audience to reconstruct the plot from the fragments of story scattered around the discourse. For example, the show *Boomtown* (2002) illustrates the same scene from multiple vantage points, allowing the audience to restructure the full story from fragments seen onscreen. Finally, the audience "plays" with narrative, constructing the full experience from a variety of sources. For example, just as the characters on *Lost* remember and piece together their lives in a variety of ways, so too does the television program *Lost* piece together the narrative through a variety of temporally displaced narrative episodes (represented to the viewer as flashbacks, flashforwards, and flash-sideways), and so too, then, does the viewer piece together the complete story, by assembling the fragments of narrative scattered across multiple characters, episodes, and narratives. At three levels, contemporary television represents fragmentation; at three levels it presents ways to (re)assemble it.

To describe these types of narratives on television, Mittell coined the term "narrative complexity."[55] For Mittell, narrative complexity is more than just a serialized plot—it defines many contemporary television narratives as a "redefinition of episodic forms under the influence of serial narration."[56] To be narratively complex, a television show must have conclusions (or resolutions) for most episodes, while also continuing a larger story arc across the series. Mittell describes shows as generically diverse as *Alias* (2001), *Veronica Mars* (2004), *Desperate Housewives* (2004), and *Lost* as ones that demonstrate both a complex narrative enigma at the center of the show as well as individual episodic plots. Audience members watch these narratively complex shows not only to solve the central "riddle," but also to explore vast textual worlds revealed episode by episode. That is, narratively complex shows establish detailed long-running stories whilst not ignoring the episodic discourse inherent in television storytelling.

Traditionally, textual research has tended to use spatial metaphors to explore key moments in the television text.[57] For example, Michel de Certeau's *The Practice of Everyday Life* sees audience members in "the space of the other [media producers]."[58] Work on convergence by Jenkins and on distributed narrative by Jill Walker describes transmediated narratives through similar spatial language: narratives may sprawl in one medium or may be dispersed among many, allowing audiences to explore different facets of narrative information.[59] Large, fan-created wikis like *Lostpedia* are one example of an audience's spatial exploration of narrative exposition.[60] As Sara Gwenllian-Jones has described, "The appeal of … vast, transmedia fictions lies precisely in their invitations to immersion and interactivity; they are constructed, marketed, and used by fans not as 'texts' to be 'read' but as cosmologies to be *entered, experienced and imaginatively interacted with.*"[61]

Yet, a spatial understanding of reception is ultimately limited, and opening up television studies to *temporal* analysis can be fruitful. Temporal displacement can create emotional affect in the audiences of the shows. As scholar Lauren Kogan persuasively illustrates, one of the key features of *Lost* that intrigues the fan base is the nonlinearity of the series: it "tries in every way possible to interweave a … sense of temporality [in]… its various clues."[62] By playing with the sense of time, the producers of *Lost* (as well as of other temporally displaced shows such as *How I Met Your Mother* and *Life on Mars*) create an aesthetically complex text which "reflects the simultaneity present in our everyday lives."[63]

Complexity of this sort cannot be completed without the appreciation of aesthetics by an active, participatory audience. As Jenkins writes, "most popular culture is shaped by a logic of emotional intensification…at its best, [it] makes us think by making us feel."[64] Gray shows that one of the key creative powers of television entertainment is its "ability to move us."[65] Spectacle, as aesthetic appreciation, is integral to television viewership. Yet, a finely tuned narrative can create an affective moment of "wow" in the audience, just as a moment of spectacle does.[66] This operational aesthetic, what Mittell sees as a show's "calling attention to the constructed nature of the narration and asking us to marvel at how the writers pulled it off," elicits spectator attention and emotional attachment.[67] The audience experiences breaks in the temporal structure of the narrative, only to "wow" at their return to the relative "present" of the show with their knowledge of the narrative somehow altered. A particularly salient example appears at the end of the third season of *Lost*. Already used to temporal shifts through two seasons' worth of flashbacks, viewers of the third season saw jumps in narrative time

(barring kinescope or telecine filming). What we once watched as "broadcasting"—that is, television programs that were aimed to appeal to a wide range of viewers—we now see as "narrowcasting"—programs aimed at niche audiences. Gone are the days when 106 million people tuned in to the finale of *M*A*S*H*.[72] Today, shows are lucky to get a tenth of that, and many are cancelled if they don't. Studying how television uses time helps us understand in greater detail this most popular art.

Notes

1 See Jane Espenson, "Playing Hard to 'Get'—How to Write Cult TV," in *The Cult TV Book*, ed. Stacey Abbott (London, UK: I. B. Taurus, 2010).

2 Henry Jenkins, "Transmedia Storytelling," *Technology Review*, 15 Jan 2003, http://www.technologyreview.com/biotech/13052/?a=f (accessed 29 Dec 2010); Henry Jenkins, *Convergence Culture: Where Old and New Media Collide* (New York: New York University Press, 2006); Jason Mittell, "Sites of Participation: Wiki Fandom and the Case of *Lostpedia*," *Transformative Works and Cultures* 3 (2009), http://journal.transformativeworks.org/index.php/twc/article/view/118/117 (accessed 01 Jan 2010); Jonathan Gray, *Show Sold Separately* (New York: New York University Press, 2010); Gillan.

3 Jenkins, "Transmedia"; Jenkins, *Convergence Culture.*

4 See Paul Booth, "Frak-tured Postmodern Lives, or How I Found Out I Was a Cylon," in *Battlestar Galactica and Philosophy*, ed. Josef Steiff and Tristan Tamblin (Peru, IL: Open Court Publishing, 2008).

5 See, for a discussion of this type of hegemonic influence on television audiences, Ien Ang, *Living Room Wars: Rethinking Media Audiences for a Postmodern World* (New York: Routledge, 1996).

6 http://www.cracked.com/video_18156_a-trailer-every-academy-award-winning-movie-ever.html.

7 Kristen Daly, "Cinema 3.0: The Interactive-Image," *Cinema Journal* 50, no. 1 (2010): 82.

8 Booth, *Digital Fandom.*

9 See Lynn Cooke, "A Visual Convergence of Print, Television, and the Internet: Charting 40 Years of Design Change in News Presentation," *New Media and Society* 7, no. 1 (2005).

10 Bolter and Grusin.

11 Caldwell, "Convergence Television," 51.

12 Anne Friedberg, "The Virtual Window," in *Rethinking Media Change: The Aesthetics of Transition*, ed. David Thorburn and Henry Jenkins (Cambridge, MA: The MIT Press, 2004), 338.

13 William Boddy, "Redefining the Home Screen: Technological Convergence as Trauma and Business Plan," in *Rethinking Media Change: The Aesthetics of Transition*, ed. David Thorburn and Henry Jenkins (Cambridge, MA: The MIT Press, 2004).

14 Gray, *Show Sold Separately.*

15 Ibid.

16 Gillan, 2.

17 David Thorburn and Henry Jenkins, "Introduction: Toward an Aesthetics of Transition," in *Rethinking Media Change: The Aesthetics of Transition,* ed. David Thorburn and Henry Jenkins (Cambridge, MA: The MIT Press, 2004), 11.

18 Fredric Jameson, *Postmodernism: Or, the Cultural Logic of Late Capitalism* (Durham, NC: Duke University Press, 1991).

19 Jean Baudrillard, "The Ecstasy of Communication," in *The Anti-Aesthetic,* ed. Hal Foster (Port Townsend, WA: Bay Press, 1983), 133.

20 Lauren Kogan, "Once or Twice upon a Time: Temporal Simultaneity and the *Lost* Phenomenon," *Film International* 4, no. 2 (2006): 50.

21 See Erlend Lavik, "Forward to the Past: The Strange Case of *The Wire,*" in *Relocating Television: Television in the Digital Context,* ed. Jostein Gripstrud (New York: Routledge, 2010).

22 Jeffrey Sconce, "What If? Charting Television's New Textual Boundaries," in *Television after TV,* ed. Lynn Spigel and Jan Olsson (Durham, NC: Duke University Press, 2004).

23 Michele White, "Television and Internet Differences by Design: Rendering Liveness, Presence, and Lived Space," *Convergence: The International Journal of Research into New Media Technologies* 12, no. 3 (2006): 342.

24 See Daly, 83.

25 See Derek Kompare, "Publishing Flow: DVD Box Sets and the Reconception of Television," *Television and New Media* 7, no. 4 (2006); Lotz; Ross.

26 David Thorburn, "Television as an Aesthetic Medium," *Critical Studies in Mass Communication* 4, no. 2 (1987): 162.

27 Ibid., 165–66.

28 Ibid.

29 Jason Mittell, "*Lost* in a Great Story: Evaluation in Narrative Television (and Television Studies)," in *Reading Lost: Perspectives on a Hit Television Show,* ed. Roberta Pearson (London: I.B. Tauris, 2009), 122.

30 Jason Mittell, "Previously On: Prime Time Serials and the Mechanics of Memory," *Just TV,* 03 July 2009, http://justtv.wordpress.com/2009/07/03/previously-on-prime-time-serials-and-the-mechanics-of-memory/ (accessed 03 Mar 2010): ¶2-3; see also, "Previously On: Prime Time Serials and the Mechanics of Memory," in *Intermediality and Storytelling,* ed. Marina Grishakova and Marie-Laure Ryan (Boston, MA: De Gruyter, 2010).

31 Shawn Shimpach, *Television in Transition: The Life and Afterlife of the Narrative Action Hero* (Oxford, UK: Wiley-Blackwell, 2010), 2.

32 See also Thorburn, "Television as an Aesthetic Medium," 163.

33 Raymond Williams, *Television: Technology and Cultural Form* (1974) (New York: Routledge, 2003).

34 See, in particular, Michael Kackman, Marnie Binfield, Matthew Thomas Payne, Allison Perlman, and Bryan Sebok, eds., *Flow TV: Television in the Age of Media Convergence* (New York: Routledge, 2011).

35 Sconce, "What If," 94.

36 Tania Modleski, "The Rhythms of Reception: Daytime Television and Women's Work," in *Regarding Television,* ed. E. Ann Kaplan (Fredricksburg, MD: The American Film Institute, 1983).

37 Boddy, "Redefining," 195; see also Jeffrey Sconce, *Haunted Media: Electronic Presence from Telegraphy to Television* (Durham, NC: Duke University Press, 2000).

38 William Boddy, "Interactive Television and Advertising Form in Contemporary U.S. Television," in *Television after TV: Essays on a Medium in Transition,* ed. Lynn Spigel and Jan Olsson (Durham, NC: Duke University Press, 2004), 120.

39 Amanda Lotz, *The Television Will Be Revolutionized* (New York: New York University Press, 2007), 34.

40 Gillan, 2.

41 John Caldwell, *Televisuality: Style, Crisis, and Authority in American Television* (Piscataway, NJ: Rutgers University Press, 1995), 5.

42 Kackman, et al.

43 Sharon Marie Ross, *Beyond the Box: Television and the Internet* (Malden, MA: Blackwell, 2008), 5.

44 Ross, 9.

45 Glen Creeber, *Serial Television: Big Drama on the Small Screen* (London: British Film Institute, 2008), 5.

46 Uricchio, "TV as Time Machine," 29.

47 Janet Murray, *Hamlet on the Holodeck: The Future of Narrative in Cyberspace* (Cambridge, MA: The MIT Press, 1997), 256.

48 Ellen Seiter, *Television and New Media Audiences* (Oxford: Oxford University Press, 1999), 116.

49 Spigel, 2, 6.

50 Jonathan Hardy, "Mapping Commercial Intertextuality: HBO's *True Blood,*" *Convergence: The International Journal of Research into New Media Technologies* 17, no. 1 (2011).

51 Gray, *Show Sold Separately.*

52 Jeff Pinkner and Joel Wyman, writers, "The Day We Died," *Fringe* (Los Angeles, CA: Fox, 2011).

53 In an intertextual moment, *The Event* here appears to reference the opening of the quintessential complex television narrative, *Lost,* with its opening-episode airplane disaster and focus on non-linear storytelling.

54 The producers of *The Event* soon shifted from this temporally displaced scheme, however, and changed the format of the show to a more-comprehendible linear storyline. Of note, the show was cancelled soon after the format change.

55 Jason Mittell, "Narrative Complexity in Contemporary American Television," *The Velvet Light Trap* 58 (2006).

56 Ibid., 32.

57 John Fiske, "The Cultural Economy of Fandom," in *The Adoring Audience*, ed. Lisa A. Lewis (London: Routledge, 1992); Henry Jenkins, *Textual Poachers* (New York: Routledge, 1992).

58 Michel de Certeau, *The Practice of Everyday Life*, trans. Steven Randall (Berkeley: University of California Press, 1984), 37.

59 Jenkins, *Convergence Culture*; Jill Walker "Distributed Narratives: Telling Stories across Networks," 2007, http://jilltxt.net/txt/Walker-AoIR-3500words.pdf (accessed 01 May 2010).

60 Booth, *Digital Fandom*; Mittell, "Sites of Participation."

61 Sara Gwenllian-Jones, "The Sex Lives of Cult Television Characters," *Screen* 43, no. 1 (2002): 84, italics mine; see also Matt Hills, *Fan Cultures* (London: Routledge, 2002).

62 Kogan, 53

63 Ibid., 50.

64 Jenkins, *Wow Climax*, 3.

65 Jonathan Gray, *Television Entertainment* (New York: Routledge, 2008), 55.

66 Sconce, "What If," 109.

67 Mittell, "Narrative Complexity," 35.

68 See also M. J. Clarke, "Lost and Mastermind Narration," *Television and New Media* 11, no. 2 (2010).

69 Henry Jenkins, *The Wow Climax: Tracing the Emotional Impact of Popular Culture* (New York: New York University Press, 2007).

70 Brian Selter, "TV Viewing Continues to Edge Up," *New York Times,* 02 Jan 2011, B1.

71 "Americans Watching More TV Than Ever; Web and Mobile Video Up Too," *Nielsonwire,* 20 May 2009, http://blog.nielsen.com/nielsenwire/online_mobile/americans-watching-more-tv-than-ever/ (accessed 01 Mar 2011).

72 Tim Arango, "Broadcast TV Faces Struggle to Stay Viable," *New York Times,* 09 Feb 2009, http://www.nytimes.com/2009/02/28/business/media/28network.html?_r=1&pagallewanted=1&hp (accessed 15 Oct 2010).

Section II: Memories

Arguably, our memory is the most coherent and stable time machine we use. We use memories to travel to the past, to envision the present, and to help stage our future. In this sense, memories—as specific moments recorded in our minds—represent the ultimate *corporeal* sense of time travel. Yet, in temporally displaced television programming, it's not just the audience's memories that matter, but the characters' as well. Usually subsumed under larger issues like narrative, authorial intent, or genre, the notion of *character* is an oft-neglected aspect of television analysis. "Character" illustrates the emotional heart of television programming and fosters audience identification. Through the representation of the memories of individual characters, contemporary television narrative uses temporal displacement to help cohere an audience's emotional attention to the onscreen character.

In these two chapters, I illustrate a duality at the heart of contemporary transgenic media: by *externalizing* our memories, we scribe our self into digital existence. Our digital photographs are posted almost as soon as we take them, and we review our immediate visual memories as soon as they have been captured. We also *internalize* the "self" that we write online by reading our own rapidly changing self-conception. We become characters in our own digital lives. Obviously, there is still a separation between our on- and off-line worlds, but the mechanics by which we deal with our online lives is rapidly becoming implemented in our offline lives as well. These mechanics are represented through temporal displacement on television.

Two

Transgenic Memories

> *In any coherent time loop, there are certain objects that are created during and exist within the time loop. One common example of such an item is the hypothetical Book from Nowhere: A man brings a copy of a book with him back in time, giving it to himself, and instructing himself to reproduce the book as faithfully as he can. The book is then published, and after its publication, the man then buys the book, gets in a time machine, and starts the cycle all over again. The book is a perfectly stable physical object that actually exists, despite the fact that it seems to come from nowhere. Less certain is whether human memory works the same way.*
>
> —Yu, How to Live Safely in a Science Fictional Universe, 113

In *How to Live Safely in a Science Fictional Universe,* protagonist Yu finds himself at a proverbial crossroads. His life remains ultimately unfulfilled: he has lived for the past ten years in a time machine, aging a decade but not moving forward in time. His only reprieve from this static journey is his ability to travel back to his own past and watch his life's events unfold. He can't change or alter these events; he can only witness the moments in his life that held meaning for him, reliving them as if he were watching his life on television. Symbolically, then, the "Book from Nowhere" represents the way Yu's life emerges from inaction: he is reactive instead of active.

Yu enacts viscerally the same type of temporal mechanics that we use mnemonically everyday to revisit *our* personal past. We don't have an external time machine, but we do have our memories and our knowledge of the past, and through them we can travel to our own histories.[1] However, even though memory for us is one-way—we only remember backwards—in Yu's science fictional universe, one can remember events of the future.

Temporally displaced television programming depicts memory as a tangible and crucial aspect of the narrative. Despite Jameson's assertion that "memory seems to play no role in television," it does seem to give us the power to see our memories for what they are: creations of our minds with shifting implications for our cultural life.[2] Memory as depicted on television complicates a show's narrative timeline. Temporal complexity forms when the viewer's recollection of what he/she sees in a narrative is revealed as another person's subjective memory of that (fictional) narrative. *Boomtown*, for instance, details the connections between the police, the criminals, the law, and the paramedics in Los Angeles. One of the characters, Detective Bobby "Fearless" Smith, has recently returned from Operation Desert Storm, during which one of his comrades was shot. The first time Smith tells this story in the pilot episode, he describes the incident but changes the ending: the audience, who has just been introduced to the character, sees the comrade survive the shooting and thinks the memory to be completely authentic. Yet, when Smith retells this incident to his partner, he admits he faked the memory's ending. Importantly, the audience witnesses both versions of the events. The disjointed temporality in the pilot of *Boomtown* creates a vehicle for experiencing all different time frames at once: we flashback to the past, flashforward to the present, and (as it was revealed) flashback to an altered past that didn't actually happen. The audience is privy to this fakery because we were fooled along with the character: the fake memory becomes an augmentation of the story even though it didn't actually occur.

In its depiction of the unreliability of memory, *Boomtown* plays on our traditional notion that memory is linear. Typically we examine the past as having already happened and the future as having yet to occur. We imagine that the past is fixed and unchangeable and we see the future as unwritten, completely open to change. Yet, *Boomtown*, as well as other shows that invoke this type of play with memory, highlights the notion that memory itself is mutable—and therefore, so are any characteristics we imagine based on those memories. The past has *happened*, insofar as events once took place; but to contextualize those events as history—to give those events meaning—is to ever-change them. In terms of the historical past, we learn to re-inscribe past events using any number of contemporary, cultural tools: archeology allows us to uncover the objects of the past; anthropology gives us the ability to interpret the role of people of the past; ethnography situates the human within cultural pasts. Yet, as classic cultural theory reminds us, it is only through the prism of *contemporary* life that we inscribe *meaning* to that past.[3] The same is true of our immediate, personal memory. The past may have happened in our life, but it's only when we give those events meaning—the first date that takes on extra significance when it turns out to be with the person whom you ultimately marry—that they become *personal history*.

As I mentioned in the last chapter, the notion that our contemporary television programming can reflect this form of indeterminacy in our subjective memory is a compelling rationale for investigating the role of traditional media in describing our conceptions about the past and future. The dual narrative structure created by the subjective viewpoints of any main characters—that is, the characters' memories—creates temporal displacement by interacting with representations of the past and future. In this way, two separate but related narrative threads function to both complicate and compound the overarching storyline. By reaching to the past, re-evaluating the present, or envisioning the future, characters in these television shows symbolize the messy function that transgenic media have in augmenting or subverting our personal memories.[4] In these temporally displaced narratives, a *mediated memory* is used rhetorically as a persuasive device not just to facilitate a greater connection between audience and character, but also to engage the audience in a more robust understanding of the relationship between mediation and human memory. I define "mediated memory" as both the representation of memory on screen and the way memory itself is shaped by the media. Indeed, media facilitate much of our contemporary memory. By visually representing human memory, television programs symbolize our contemporary culture's reliance on transgenic media as constitutive of our own memories.

Mediated memory symbolizes the variability of human memory. When television narratives illustrate the *failure* of mediated memory—that is, the self-conscious perception that faulty memory as represented is, itself, faked—they thus extol the *capabilities* of unmediated human memory. The double negative reifies human memory. The device most characteristic of a mediated memory is the subjective flashback, wherein the narrative of a television program shifts to represent a character's particular viewpoint on something that happened in the past. Mittell describes these "subjective flashbacks" as "glimpses of character's memories via flashbacks [as] a common cue to trigger a viewer's own memories."[5] Subjective flashbacks may or may not be unreliable. Although flashbacks have been common on television for nearly its entire history, what has changed in the past decade is a narrative *reliance* on faulty, deceptive, false or otherwise imperfect representations.

The use of mediated memory as a temporal device symbolizes the ephemerality of transgenic media. We represent our memory in bits, and reflect on ourselves online. Transgenic media encompass a particular dialectic between the *externalization* of memory and the *internalization* of the self. Reading through my own *external* memories as I've posted them, I can form an *internal* expression of myself. In a Goffman-esque manner, digital memories become a performance for/to the self, illustrating not just how I used to behave, or what I used to be like, but also what I could possibly become again.[6] Transgenic media become a tool for users to inscribe their own memories as projections. And when transgenic media present alternate viewpoints from the past, individuals can review and re-inscribe meaning into the subjective moments from their life.

We scrutinize our office party photographs the instant we take them, we watch our history on YouTube, and we write our thoughts and opinions on public blogs. Who amongst us hasn't retaken a digital photograph if our smile looks fake, crafting the perfect memory? With our digital photographs, we can instantly see memories of our vacations, weddings, or birthdays—as they happen. Further, we can remember them again, externalizing the experience with our friends and families. The much-mocked torturous vacation photo slideshow, depicted meaningfully across a variety of sitcoms, has morphed in recent years to the omnipresent Facebook slide show, which is now demanded by pushy parents and jealous lovers. The status we post on Facebook, the updates on Twitter, the ever-expanding blog diary: these technologies help us digitize, store, and organize our memories into taxonomic, folksonomic, and ontological categories. Nothing represents this better than the ever-present threat of being tagged in a photograph from the past. My high school friend can post the most embarrassing moment in my life—a memory I would sooner forget than anything else—and there my memory resides online for everyone's amusement. Further, at any time, I can log on to Facebook and find information from the earliest days of my profile. The events of my life are laid out in a personalized narrative, open for digital expression. One consequence of this publicity is, as scholars like Zittrain and Palfrey have noted, changes in social mores and practices in concerns about privacy.[7] It's not that privacy issues have become less important to users, but that the very concept of what "privacy" encompasses may be shifting. Younger users may simply not think that some information needs (or should) be private, especially information that older generations may have once never discussed with anyone outside their family.

Whatever the rationale behind the changes in the way younger users of the web understand privacy, there is indeed a shift in the way such information is processed. Indeed, the web is becoming more important not only in terms of its representational qualities, but also in terms of its immediate pedagogical value. For example, think of the outrage felt when online social network site Friendster announced in April 2011 that it was deleting personal information from the past. Despite the fact few people still used Friendster, according to the *New York Times*, the collection of personal information from years past "turned into a time capsule with snapshots of who they [users] once were. It is a version of their history that is not in a scrapbook or dusty shoebox but is live on the Web."[8] We learn about ourselves through our own self-representations online.

And the number of these representations is growing. As of January 2011, the population on Facebook exceeded the population of North America. According to the Internet World Statistics, over three quarters of the US population use the Internet regularly; and the Pew Internet and America Life project shows that 66% of online adults use the web socially.[9] Although this may represent only slightly more than half of the US population, this number is growing. Given the rapid proliferation of transgenic media in our lives, it is no wonder that our traditional media forms attempt to attach themselves aesthetically and textually to this media growth. By representing memory as a tangible entity, by reflecting on the unpredictability and instability of memory, and by allowing for alterations in memory temporality, shows with temporal displacement like *Doctor Who, Dollhouse, Flashforward, How I Met Your Mother,* and *Lost* become emblematic of larger cultural changes engendered by our technological use, and teach us how to assemble and make sense of these ephemeral, external memories.

Mediated Memories

We live in an age of memorial and an age of instantaneousness. Our immediate, personal media use demonstrates an immense capacity for us to *remember*. What was that webpage I visited thirty days ago? Google knows. What was I feeling when I was dumped in 2005? MySpace knows. But we have also never had a more ephemeral relationship with our mediated memory use than we do now. We live in the *now*, constantly updating our statuses, continuing conversations in texts long after we've forgotten how they started. We may have access to our entire Internet browsing history, but unless we're deleting our history from the prying eyes of our bosses, spouses, or parents, how many of us spend time revisiting our online history in depth?

And yet, recording our thoughts is not just a way of reaching out to others, but also a way of connecting to ourselves. Foucault has argued that by sketching out the "different ways in our culture that humans develop knowledge about themselves" we can examine the evolution of the ethical self.[10] There are particular technologies and practices that humans have used to constitute their selves, including writing, speech, and prayer.[11] In the Hellenistic age, for instance, the act of writing was seen as a crucial tool for "taking care of oneself," as "the self is something to write about, a theme or object (subject) of writing activity."[12]

Just as the Greeks used the novel technology of writing to represent themselves, so too do we use our new technology of transgenic media to write ourselves. Memory is, of course, one of the central tenets of what makes us human: as Cowan points out, the very human connection between "emotional and empathic responses [are] linked to memory development."[13] This memory/emotion connect has been a common theme in science fiction. In the film *Blade Runner* (1982), for example, Deckard is able to identify the Replicants by their faked emotional response to their memories, a trope examined as well in 2005's *Battlestar Galactica.*[14] Artificial memories differentiate the *human* from the *machine.*

Yet, the *concept* of memory has played into the cultural meaning of media from the earliest days of mediated communication, and many philosophers have connected the use of media to the destruction of memory.[15] For Plato, the mediation of writing saps the mind of its powers of memory and concentration. Describing Socrates' disapproval of the invention of writing in *Phaedrus,* Plato argues that there will be a resulting loss in memory, wisdom and truth:

> this invention will produce forgetfulness in the souls of those who have learned it. They will not need to exercise their memories, being able to rely on what is written, calling things to mind no longer from within themselves by their own unaided powers, but under the stimulus of external marks that are alien to themselves. So it's not a recipe for memory, but for reminding...they'll entertain the delusion that they have wide knowledge, while they are, in fact, for the most part incapable of real judgment. They will also be difficult to get on with since they will have become wise merely in their own conceit, not genuinely so.[16]

In Plato's estimation, writing is detrimental to memory, and forces humans to relinquish some of their innate talents in favor of technological augmentation.[17] Plato, of course, is writing at a time of great cultural change, and the irony is that Plato's great criticism of writing is recorded for posterity *in writing*. Yet, as Illich and Sanders describe, it's not until the invention of writing that we even have a *conception* of memory in our lives. In their determination, writing *invented*, rather than *destroyed*, memory:

> only after it had become possible to fix the flow of speech in phonetic transcription did the idea emerge that knowledge—information—could be held in the mind as a store. Today, we take this idea so completely for granted that it is hard for us to reconstruct an age when recollection was not conceived [of]...it is not clear that a purely oral tradition knows no division between recollecting and doing.[18]

As deterministic a prognostication as Plato's disparagement of writing may be, with any monumental change in communication, as Walter Ong has shown, changes in the patterns of human thought do emerge. For Ong, the shift from an oral culture to a print culture brought with it a shift from a world of *sound* to a world of *sight*.[19] Yes, humanity may have lost some of its inherent skills in memorization, but we have also gained skills in the archive of knowledge and the organization of thought. There have been a number of similarly great epochal changes in the history of mediated communication: the printing press, the telegraph, and the Internet are three salient examples. Yet, with each new technology comes new repression and antipathy from those wedded to traditional modes of communication. The advent of the Internet, in fact, brought about new criticism of memory's relationship to technology; perhaps because, as Welch shows, there is a new "electric rhetoric" which finds oral traits manifest in digital conversations.[20]

Indeed, we can never escape the issues that Plato explores. Transgenic media suffer from the same stigma in terms of the loss of memory as did print. For example, in *Alone Together*, Sherry Turkle focuses on MyLifeBits, a research project from Microsoft by computer pioneer Gordon Bell, whose aim has been to create a "complete life archive" of himself, including "books, cards, letters, memos, posters, photographs, and even the logos from his coffee mug and T-shirt collection...home movies, videotaped lectures, and voice lectures."[21] For the project he recorded every conversation, photographed every meeting, saved every web browse. MyLifeBits is, as Bell puts it in Turkle's book, a way to "tell ... life stories to [your] descendants."[22] Turkle calls this form of archiving "the human desire for a kind of immortality, the ancient fantasy of cheating death."[23] Like Plato, though, she quickly questions whether "the fact of the archive [will] convince us that the work of remembering is already done."[24] Bell's project augments memory, but by critiquing MyLifeBits as a technological *replacement* of memory instead of a technological *augmentation* of memory, Turkle appears to argue for a return to a less-mediated life, a Platonic ideal of technology and memory less intertwined.

Bell's MyLifeBits project is reminiscent of another "memory extension" program—Vannevar Bush's Memex device. The Memex, a device designed to store all of humanity's knowledge in one location, and through which researchers could create stable "links" between relevant pieces of data, was conceived of during World War II to aid scientific research at a time when the scope and range of scientific data was becoming overwhelming. Although never actually constructed, the Memex is arguably one of the key technologies that conceptually helped shape the modern Internet. Bush argued that researchers were "staggered by the findings and conclusions of thousands of other workers—conclusions which he [/she] cannot find time to grasp, much less remember."[25] For Bush, the Memex *extends*, not *replaces*, "limited [human] memory."[26] Turkle's depiction of the loss of memory is more dire than Bush portends: she asks whether such memory replacement will make us "clean of remembrance? Clean of messy, unreliable associations? Do we want to be 'clean' in this way?"[27] But is this cleanliness, or merely orderliness? Transgenic media can be seen as organizational tools that many people find easier and more enjoyable to use than other media. Yet, their ubiquity and popularity should not damn them.

Further linking technology and memory, Cliff Atkinson illustrates the intimate connection between human memory and Twitter as a technology of personal communication. Like memory, he argues, Twitter organizes in an "information hierarchy"; that is, "we have to figure out what's most important to understand it."[28] Memory works in the same way: "since we can't possibly access all the information in our minds at one time, we access the top-level information we need and then deeper levels of detail when we need it."[29] Memory is not just a tool for Atkinson, but also a heuristic device. The same type of heuristic is pointed out by narratologist Marie-Laure Ryan, who describes how the immersion of the reader in a hypertext narrative can draw on the mnemonic techniques of the "memory palace." A tool used for memorization, the memory palace teaches readers to construct a mental visage of a house and situate all details of the text in rooms in the palace.[30] By wandering through this mental construction, readers perform a detailed recall of mediated information. Media, in this case, help bolster the reader's memories. Additionally, Ekaterina Haskins argues that the Internet has fostered a new form of memory creation and acquisition. In her discussion of *The September 11 Digital Archive*,[31] an online repository for collecting, preserving, and presenting the history and memory of the September 11th terrorist attacks, she notes how a "'digital memory,' more than any other form of mediation, collapses the assumed distinction between modern 'archival' memory and traditional 'lived' memory by combining the function of storage and ordering on the one hand, and of presence and interactivity on the other."[32] For Haskins, the Internet allows private memory to become collective memory, heralding a shift in complex representations of memorials from individualized to communal.

The effects of the externalization/internalization dynamic of memory representation in transgenic media are perhaps most vividly felt with online videos. Michael Strangelove calls this a "hidden paradox" of digital technology. We record memories in digital and collect them on our home computer and yet, as Strangelove argues, we lack a formal, coherent organizational system for the influx of ever more storage/memory.[33] As he suggests, therefore, one of the consequences of YouTube "is that it is expanding the memory of the citizenry."[34] Indeed, one person's memory, augmented by YouTube's storage capacity, can bolster another person's as well, as memory sparks memory: Strangelove describes the case of the *Disneyland Home Movies*, a series of films from the late 1960s/early 1970s and posted to YouTube. Garnering over 65,000 views, the home movies sparked other people's memories, as this comment describes: "Thanks for sharing this video! It brings back great memories…"[35] The transgenic media here have not damaged the ability of its users to hold memories, but rather have fostered those memories to reemerge.

Memory and Television

As a medium, television plays a central role in the way we experience, create, and bolster memory. One critical aspect of memory on television is the way televisual texts foster audience recall. As Mittell has explored, contemporary television programming has used the notion of seriality to "trigger, confound, and play" strategically with viewers' memories in an effort to "engage viewers and enable long-term comprehension."[36] Narratively complex television shows use serial storytelling to remind short-term viewers of particularly relevant narrative information, and to appeal to long-term viewers who may be more knowledgeable. In order to court new viewers, producers often feel it necessary to repeat information across a range of shows through what Mittell calls "diegetic retelling." So, characters on television programming will often refer to each other by name much more often than we do in real life, and they might mention titles and job descriptions more often than one would think "natural" in non-mediated human conversation. This diegetic repetition functions as a way of reminding viewers who characters are and what their roles are in the narrative, so that if a viewer misses an episode or two, or even takes a year-long break, he or she can catch up quickly.

Mittell describes a number of formal strategies that television shows present to manage this viewer comprehension, including aspects that I have termed temporal displacement.[37] To understand a narrative it is therefore necessary to mentally "map" the unfolding of the narrative, as Altman describes, which calls on "our memory of the text at hand."[38] The complexity of today's television programs, however, has led to more oblique redundancies and repetition. Rather than directly referencing past events, a character's diegetic references might seem more natural, more realistic. This naturalism is a key aspect of what Espenson calls the participatory aspect of television narrative.[39] In this way, the memory of the *viewer* becomes linked with the memory of the *characters*, as characters are reminded in similar fashion as is the audience. For example, in "Day One" of *Torchwood: Children of Earth*, the third season of this 2008 BBC show, protagonist Gwen Cooper walks into her station at Torchwood and brushes her fingers past a photograph of two people. For those that watched the first two seasons, this small gesture has enormous weight, as the two people depicted in the picture are Gwen's recently deceased colleagues Owen and Tosh. But for those new to the series, the moment would have held little meaning except to build Gwen's character traits.

The way memory is presented in the media focuses attention on the use of memory as a natural, human device. But by problematizing memory, by playing with different views of "memory" on screen through flashbacks and flashforwards, contemporary television programming offers different "takes" on the way memory can function in a digital society. The failure of a mediated memory symbolically turns to affirm the specificity and reliability of corporeal memory. Specifically, I want to examine three different ways that mediated memory appears to fail on television as a device of temporal displacement: through flashbacks, we see *fake memories* and *created memories*, and through flashforwards we see *future memories*. Given the importance of memory within our quotidian media use, the attention of television programming on flawed or false representations of memory can tell us much about our human reliance on memory. In some of these programs, the idea of memory as an *external* device for recording or changing appearance becomes obvious through narrative devices and character activities. Other programs represent the idea of memory as an *internal* device which permits changes in conceptions of the self. In both cases, memory serves as a unifying device that structures and orders contemporary mediated life.

Television Representation of Transgenic Media Memory

The representation of character memory in the text of the television program engages a temporally complex narrative, as it refers back to different points in time within each episode and complicates the linearity of the storyline. Memory becomes a tool to portray a delicate balance between the interpretation of a narrative and the composition of a complex narrative structure. A focus on memory presented by each of these television programs presents a symbolic connection between viewers and shows, a connection that viewers take with them as they use other media. The *nature* of one mediated interaction (television<—>viewer) symbolizes or affects the *nature* of another interaction (transgenic media <—>user).

The television program demonstrates memory by showing onscreen, either through flashbacks or flashforwards, narrative digressions in time. A standard of television and literature for years, the flashback presents onscreen a character's previous life events as if they were represented for the viewer. Contemporary television seems to rely more and more on flashbacks as an integral part of a show's structure. However, important differences exist between the representations of those memories in contemporary complex television compared to how they've been conceptualized in the past. Traditionally, television flashbacks have mirrored the typical Hollywood flashback: it "is typically not consistently 'subjective'; often it tell us things that the narrating or recollecting character could not even know."[40] In contrast, flashbacks in contemporary television often take the viewer into a representation of the character's subjective experiences. Viewers may be able to piece together their lives because of what they are told, but every new revelation of memory helps construct a "whole" but necessarily incomplete character. As Eco shows, "fictional characters live in an incomplete … world which necessarily depends on the viewer to create."[41] The television program merely shows the events; viewers must piece everything together after seeing that mediated memory.

Flashback as Memories

Temporal displacement can function in a number of different ways onscreen. In some television shows, for example, the entire premise is based on an extensive flashback, or one in which the entire narrative is framed as a flashback. Mittell describes these extensive flashbacks as "alterations in chronology," which serve to "to frame an entire episode's action in the past tense."[42] Shows that fit this mold include *Oliver Beene* (2003), *The Black Donnellys* (2007), *The Nine* (2007), and *How I Met Your Mother*. This effect may increase the audience's aesthetic appreciation for narratives' twists, turns, and oddities by focusing on the concept of the narrator, often accomplished by means of a voiceover. Like its predecessor (and obvious influence) *The Wonder Years* (1988), the television show *Oliver Beene*, for example, is narrated from the point of view of one character relating his memories of the past. Other shows feature additional nonlinear alterations in the plot within extensive flashbacks, as in the show *Jack and Bobby* (2004). The titular characters are two brothers whose story is punctuated (and paralleled) by a number of interviews and reminiscences by characters from the shows' diegetic future. The first episode tells the audience at the onset that one of the characters will grow up to become President; the enjoyment of the episode lies not just in the unfolding of the narrative, but also in figuring out which of the brothers would become Commander in Chief.

In narrative terms, the focus on the narrator is a central theme to any extensive flashback, and as Bal shows, "examining the temporal aspect of the narrating requires one already to take into account the 'person' of the narrator."[43] For example, the narrator of *The Black Donnellys* retells the events in the lives of the brothers from his jail cell. Early in the show's run, this narrator is revealed as unreliable: he forgets details, revises history, and tells the same story with different outcomes.[44] In the world of television temporality, the moments when the narrator shows unreliability are moments when the audience must face the realization that everything they are watching might be suspect; might be a fiction. The attuned viewer exclaims surprise, rethinks the past episodes, and thus alters his/her view of the story itself. The purpose of the temporally displaced narrator, therefore, is to force the television show into a new storytelling mode: one predicated on meta-moments and the hypermediality of the narrative.

As a case in point, *How I Met Your Mother* uses multiple flashbacks to tell a temporally complex story. Almost every episode of *How I Met Your Mother* opens with a two-shot of two children and the on-screen title "2030"—indicating that the show takes place almost twenty years in the audience's future. The first episode of the series, "Pilot," depicts the eponymous narrative situation: the off-screen narrator, talking to the children, tells them that they will hear the story of how he met their mother. Occasionally, the narrative cuts back to these two children and their father, who comment on the unfolding (past) plot. In this way, the show seemingly tells two stories: the first is the titular one of how the main character (Ted) finds and falls in love with his spouse; the second story is the narration of the first story to the children and their reaction to that story.

The flashback structure of *How I Met Your Mother* can be an extremely effective vehicle for extended comedic reveals. One pertinent example reveals a "Telescoping Flashback"—telling a story entirely through flashbacks within flashbacks.[45] In the third season episode "How I Met Everyone Else," each main character recounts the story of how each met the others: humorously filling in the gaps of other people's memories through their own recollection. Each story told by a character is illustrated onscreen with a flashback, but of course the entire show is also a flashback. This type of flashback-within-a-flashback is common in *How I Met Your Mother*, and we can also see it in the episode "Return of the Shirt." Protagonist Ted decides to get back together with Natalie, an ex-girlfriend he had dumped quite painfully three years earlier. 2030 Ted relates the story of 2007 Ted telling this story about 2004 Ted to his friends at his local bar, McLaren's. The first flashback-within-a-flashback occurs when 2007 Ted initially relates the breakup to his friends, and the show flashes back to Ted on the phone. Later, the flashback continues and the audience learns that Ted's phone call is to an answering machine. His obviously flawed interpersonal skills become the butt of the joke, not because the audience is told of his heinous breakup mistake (never leave a message!) but because the temporal order of the episode has been disrupted, causing a jump in time which jolts the audience into different "presents." Finally, the reveal of the answering machine itself becomes more extensive with a third flashback, when the focalization of the flashback changes and Natalie herself finishes the story: Ted's message on the answering machine was left just seconds before Natalie arrives home for a surprise birthday party, and a shocked group of people listen as Ted breaks up with her via the message. What *How I Met Your Mother* effectively portrays with this final flashback-within-a-flashback-within-a-flashback is a fluctuating narration within temporal displacement. Even though Natalie "completes" the flashback, the aesthetics of the past scene and the content of the flashback remain identical to when Ted also flashed back to this time period. The change in the point of view, therefore, augments the complex storytelling of *How I Met Your Mother*.

Fake Memories

The use of *fake memories* occurs when television programming displays a character's memory and then articulates the inaccuracy of that memory, "faking out" the audience. An exemplary fake memory can be seen in the *Community* episode "Paradigms of Human Memory," during which all the main characters have flashbacks to moments that have never appeared in the show. In a sense, "Paradigms of Human Memory" is a clip show of never-before-seen clips.[46] For example, the septet of characters (Jeff, Annie, Abed, Troy, Shirley, Britta and Pierce) all flashback to an "episode" where they were all trapped in a Wild West-esque adventure, complete with stock elements like a racist prospector and a newfangled locomotive. Each of these characters "remembers" these scenes, but each of the scenes is completely new to the audience. Other minor characters diegetically refer to these events as well, playing on the audience's (lack of) memory.

Returning to *How I Met Your Mother*, we can see another example of this fake memory flashback. As we've seen, the entire show is a flashback, but within this framework exist more flashbacks in each episode. In the episode "Murtaugh," Marshall recounts his attempt at coaching a kindergarten basketball team. As he describes the trials of his team, the episode breaks with the temporal structure of the narrative to illustrate his memory. During the beginning of his exaggerated story, the team facing his kindergarteners consisted of professional basketball players. Marshall then tells the same story, but this time includes a werewolf opponent. Finally, the flashback shows several shots of the werewolf slam-dunking the basketball, in an obvious homage to the classic teen comedy *Teen Wolf*. The continued temporal displacement of the re-telling of the narrative builds the audience's expectations, which are eventually rewarded with the visual quotation of *Teen Wolf*, a move which both satisfies the building anticipation of the joke while at the same time eliciting a direct affirmation of the false mediated memory of the show.

The telescoping flashback structure of *How I Met Your Mother* offers another argument for a subjective understanding of mediated memory. In "The Mermaid Theory," as Future Ted narrates the story of Lily's discussion of her pregnancy with Barney, and Marshall's budding friendship with Robin, it becomes obvious that his memory is failing him. For example, Future Ted remarks that Lily was about to insult Barney, but Future Ted can't quite recall what she said—an unusual occurrence for someone who apparently has a phenomenal, telegenic memory. As Future Ted fumbles to remember the insult, Lily and Barney break the fourth wall by looking directly at the camera, inviting the audience to identify with the characters' annoyance at this break in memory. The rash of memory-mishaps continues throughout the episode, as Barney later echoes Future Ted's (lack of) memory, stating "I'm still mad at you for ... *something*...and that *other thing*," phrases that are deliberately left vague because Future Ted can't remember what irked Barney. One key scene that continues to crop up in the fake memory of Future Ted takes place between Barney and Lily on the street in front of McLaren's Pub. Future Ted describes how Lily and Barney eventually make up when Lily saves Barney's ice cream cone from being spilled after Barney is almost hit by a motorcycle, and Barney magically floats a beer bottle in front of Lily.

It turns out that Future Ted was confused in his memories, and had swapped elements from two different time frames in his story. Both time frames, however, were depicted onscreen at the same time, mashing up the elements into a (semi)-coherent whole. Barney and Lily were having an argument, but it took place after the point in time Future Ted thought it did. In that future time (past time to Future Ted, however), Lily was pregnant; the magically floating beer was actually balanced on her bulbous belly; and Barney saved Lily from a speeding motorcycle (no ice cream, it turns out, was involved—that was also a false memory).

Ultimately, the fake memories depicted in "The Mermaid Theory" represent in miniature the temporal displacements occurring all over contemporary television. Two different time frames occurring at the same time onscreen mashup to construct one coherent timeline. At the same time, using the knowledge of the audience, the writers of the show construct a visual representation of these faked memories, which reasserts the power of *human* memory over *mediated memory*. Future Ted can't remember the events correctly, but because his vision is mediated through the television screen, the audience's fear over the memory mistake displaces towards the technology.

Television reflects today's transgenic media use. With digital technology comes not only the power to record but also the power to alter digital memory. Through the manipulation of digital data, images, videos or files, one can effectively change the past, or at least our representation of it. And the more our physical memories rely on the digital ones, the more such changes can have tangible effects on the way we live our lives. In fact, to replay these "faked" memories across the cultural landscape means conforming our physical existence to digital temporality. Just as we see the results when Future Ted searches his memory in order to realize the existence of a truthful past, and just as we see Detective Smith retell his story in order that the ending become truthful, so too must we all experience our transgenic media in a space where the fluidities between "truthfulness" and "fakery" collide.

To Photoshop our memoires, however, is not just to replicate the very fears that Plato described two and a half millennia ago, but also to reaffirm the power of the individual to shape his/her own memories as they occur. To displace memory temporality in a television program is to display different time frames as a character remembers them. The act of accidentally mis-remembering, as Future Ted does, or of deliberately mis-remembering, like Detective Smith does, is also an act of *creation*, as the new memories have created new onscreen representations. Such creativity may be antithetical to authentic memories, a judgment that becomes negatively weighed in a culture which values truth and authenticity above most else. But the creation of memory is an important tool in the way we organize our cultural life: how we choose to represent the past tells us more about ourselves than it does about that represented subject.

Created Memories

Another method of temporal displacement in contemporary television narrative programming is the use of *created memories* to alter the depiction of character traits. An external agent deliberately constructs a memory and implants it in a person to change the character who now holds that memory. For example, in *Battlestar Galactica* the Cylon sleeper agents are unknowingly implanted with (fake, created) memories of their lives as humans. Like the discovery of the Replicants in *Blade Runner*, these created memories of the Cylons become one of the keys to discovering who the "final five" are towards the end of the series. An exemplary case of created memories is Joss Whedon's *Dollhouse*. In the show, characters' memories and personalities are downloaded out of their bodies and replaced by other memories/personalities. These created memories (re)present the replacement of "authentic" memories with new ones. The memories implanted in the Dolls affect the personality of the character: given the memories of a kickboxer, a doll could then "know" the skills of kickboxing. Wealthy business people can special-order particular personalities to be placed in young, attractive bodies for a specified amount of time. The Doll is a complete realization of the new personality—no trace of the old is supposed to surface—and thus subscribes to a true Cartesian dualism. Yet, the audience experiences two timelines at once: the timeline of the Doll with the created memories, and the timeline of the Doll's original memory, helpfully kept in storage for future use.[47]

Relevantly, the long-running *Doctor Who* has presented a view of created memory as both complementary to and characteristic of character traits and mediated narrative.[48] A recent emphasis in the program is a concentration on the power of the mind and memory to shape the reality of the characters. For example, in the 2010 Christmas special "A Christmas Carol," the Doctor changes the memory of Kazran Sardick, in order to physically change him as a person. In a takeoff of Dickens' classic *A Christmas Carol*, the Doctor plays ghosts of Christmas Past, Present, and Future to Sardick's Scrooge. Sardick, a miserly old man who rules over a planet infested with heavy fog and flying sharks, refuses to allow the people working on the planet to take a break and have a Christmas celebration. Furthermore, a spaceship filled with innocent people has broken down and is hurtling towards the planet and certain doom. Sardick has the power to save these people, simply by turning off the barrier that surrounds the planet and allowing the ship to glide to safety, but he refuses: why waste time and energy saving a few hundred people, he reasons?

Enter the Doctor. Unable to turn off the energy barrier himself (the controls are tuned to Sardick's brainwaves alone), the Doctor resorts to Plan B: change Sardick's memories to effect an emotional alteration in the Scrooge-like Sardick. To change these memories, the Doctor—an experienced, if erratic, time traveler—opts to travel back in time to when Sardick was just a boy and discover why he has become so stubborn and obstinate. Yet, to do so, the Doctor doesn't just travel back in time, he also *shows* Sardick these changes via Sardick's own home movie collection—a second-order visual representation which affects not only Sardick but also the audience.

First the Doctor travels to Sardick's past and interacts with him as a child. Sardick (the man), who is watching old home movies, sees this interaction occur on the screen; the time lines become merged as the past, depicted onscreen, affects the present (similarly depicted on the audience's television screen). Imagine watching your childhood videos on YouTube, but seeing them change as your memory is literally created in front of your eyes. The Doctor then travels to Sardick's present, with the intention of seeing how his meddling in the past affected the man in the present. When he learns that his interference did not produce the results he desired, the Doctor brings the boy Sardick from the past to the elder Sardick's present—the Doctor becomes the Ghost of Christmas Future to Sardick's past childhood in Sardick's present life. When the boy sees what he was to become in his own future, the man responds by changing his mind: not because he saw himself from the past, but because he saw himself in the future. The change is not just mental, however: for in Sardick's case, his memory (like the Dolls in *Dollhouse*) is changed so completely he literally becomes a different person. The alteration in his memory forces a physical shift as well, and his brainwaves literally change: he is unable to adjust the controls, tuned as they are to his previous physical existence. Thankfully, due to the Doctor's *Deus ex Sonic Screwdriver*, the passengers on the plummeting craft survive the landing.

Doctor Who contributes to this contemporary temporal displacement by representing two different modes of memory operating at the same time onscreen. First, the original memory of Sardick reveals itself through the home movies that he has constructed; and second, the Doctor travels to the making of those movies to illustrate how changing the memory becomes influential in Sardick's physical/mental makeup. Importantly, the audience sees both memories *at the same time*, displacing the temporal situation and creating a dual-mediated narrative: the memory is shown to the audience simultaneously on the television screen and on the screen in Sardick's mansion. The narrative focus, shifting from one medium to the other, displaces the audience within the diegesis of the television story.

Ultimately, the creation of memory on television represents a similar track within transgenic media. Just as the Doctor can alter the personality of Sardick, or as the Dolls in *Dollhouse* can be different "people" with the creation of new memories, so too can digital technology (re)present new memories through the digital manipulation of images, video, and audio. We can't change the past, but we can change our representation of it. On *Lost*, for example, the character Desmond carries with him the last reminder he owns of his fiancé, Penny: a photograph of the two of them with Honolulu in the background. Viewers see the photograph a few times before they flashback to the actual taking of the photograph, when it is revealed that the Honolulu background was just a backdrop screen owned by a street photographer in London. The trickery of the photograph—assuming one location yet representing another—mirrors how digital technology could be altered and changed via digital manipulation of the memories we have of locations and times in the past. Further, in a meta-textual moment, in the original episode in which the image of the photograph appears, "Orientation," the woman in the picture isn't Sonya Walger, the actress who plays Penny, but rather is played by an unnamed actress (the importance of the character had not yet been established). In later airings (and on the DVD), once that relationship had been established, the producers digitally altered the original photograph with a version that had itself been digitally altered.[49] Of course, altering the mementos of the past is not a new practice: photographic trickery goes back as far as, at least, the Cottingly Fairies. Yet, never before has it been so easy or so ubiquitous in our culture as it is now.

Flashforward as Memory

Although flashbacks have always been a part of television textuality, it is only recently that the flashforward has emerged as a tangible aspect of televisual content. While a flashback depicts an event from the past, the flashforward depicts events from the future. As Bordwell has shown, the flashforward "lets us glimpse the outcome before we have grasped all the causal chains that lead up to it."[50] A commonly used flashforward technique in cinema details the culmination of the lives of the characters written on screen at the end of the film. "Curt Henderson," the title card reads in *American Graffiti* (1973), "is [now] a writer living in Canada." Yet, on television, the recent spate of flashforward elements not only *tells* but also *shows* these future events. These *future memories* record what is to happen, or what might happen in the future, in temporally complex narratives. For example, *Lost* uses future memories in the second half of the series to depict what happens to the characters after they make it off the island (and as they then want to return). Glimpses of the characters on the island are juxtaposed with their adventures in the "future." The future memories mean that the narrative goal of the show changes—instead of seeing the characters get off the island, the goal becomes seeing why they want to return, or how their experiences on the island have changed them.

Future Memories

The presentation of *future memories* in *Flashforward,* for example, allows for the development of temporal displacement. The temporally displaced narrative—the audience sees pieces of future events—not only makes obvious the narration, but also forces the narrative arc into a binary adventure: will these visions come true or not? The narrative questions, the queries that keep the audience of this serial narrative guessing, force a reconceptualization of the role of the "answer" in long-form narrative. In *Flashforward,* as in the second half of *Lost,* the audience already knows the answers; the narrative becomes an understanding of the questions.

The new *Doctor Who* series also uses a form of future memory. At the end of "The Pandorica Opens," for example, two events seem to happen simultaneously: the Doctor being put into the Pandorica and companion River Song trying to stop the Doctor's ship, the TARDIS, from exploding. Yet, the seeming-simultaneity of the events indicates that history and memory are both multi-linear. As he stands in the year AD 102 in front of the Alliance of alien ships—Dalek, Cybermen, Sontaran, Judoon, Nestene, Auton, Slitheen, Sycorax, Zygon, and more—the Doctor dares them to approach: "Remember every black day I ever stopped you," he cries, "and then, do the smart thing. Let somebody else try first!" Yet, the victories he refers to happen in the future. The Doctor's notion of "memory" is thus evoked here in an odd way, as the uni-directionality of memory (we remember the past in the present, but we have no memories of the future) reverses: the Doctor is counting on *future* memories having equal weight to *past* ones. The ships recede precisely because their memory of the Doctor comes both from their future and from their past.

In fact, memory itself is key to the Pandorica, as it is constructed from his companion Amy Pond's memories, the Alliance tricks the Doctor by influencing his memories of Amy Pond, and in the end it is Amy's memories of the Doctor that save the universe. Yet, the event that the alien Alliance is trying to prevent—the explosion they all "remember"—is occurring *at the same time.* The Alliance's attempt to stop the explosion is what triggered it in the first place. They create memories in an attempt to prevent them. Their memory is triggered by the Doctor's very attempt at uncovering the truth. The Pandorica becomes Yu's proverbial "Book from Nowhere," a case of, as time travel scholar Paul Nahin states, the *simultaneous causation* at the heart of a causal loop.[51]

Flashforwards also appear in *How I Met Your Mother*. Indeed, the flashback-within-a-flashback structure already in place inherently presents a flashforward of sorts—any time Future Ted indicates a past event which takes place chronologically after the contemporaneous events on the show, it's a flashforward disguised as a flashback. Yet, even in this complicated structure there are still more recognizable flashforwards. In the second episode of the series, "Purple Giraffe," 2007 Ted relates his plans to woo new crush Robin at a party. As he describes his actions which will take place in the future, the temporal location of the scene changes to represent Ted at this party. As 2007 Ted utters dialogue he is planning to say to Robin in voice over, the characters in the flashforward mouth the same words, linking visually and auditorily the two scenes. At one point, the characters in the flashforward look at the camera in disbelief at something 2007 Ted says, linking the flashforward to the past.

Future memories can also come in the form of a mediated communication for the characters in the show. In the first season of *Heroes*, time-traveling protagonist Hiro learns about the plot to destroy New York through reading a series of comic books that foretell the future. The characters in *Heroes* have been endowed with super-human abilities: Hiro can time travel, Claire can regenerate body parts (and is seemingly immortal), and Nathan can fly. The comics that Hiro reads, *9th Wonders!*, was written by Isaac Mendez, who had the power to draw the future. In the episode "Don't Look Back," Hiro sees himself on the cover of issue #14, depicting his journey to New York City—which he undertook in the first episode of the series.

Television shows use extensive flashbacks and flashforwards to confuse the temporal location of the narrative. The pleasure of the text lies not in a readerly aspect of the text, in which the audience passively enjoys the pleasures of the text, but rather in a writerly aspect of the text, as the audience (re)-constructs the temporal situation. [52] For the viewer this reconstruction factors into the schizophrenic notion of cohesive time in contemporary society: by structuring an unreliable narrator, the creators of temporally displaced narratives offer viewers a window to their own temporally displaced lives, and a coherent set of mechanisms to piece them together.

Conclusion

These types of abnormal memory representation do not necessarily function separately; some television programs utilize combinations to tell the fractured narrative. In *30 Rock*'s "I Heart Connecticut" episode, for example, Pete Hornberger challenges co-worker Frank to an arm wrestling competition. To everyone's surprise, the meek Pete triumphs—and to demonstrate how he has amassed such upper body strength, the scene quickly cuts to a flashback of Pete's memory of batting children away with his arm. The rest of the episode follows Pete as he beats every co-worker in the office, until he faces the team's "bully," a burly man named Reggie. During the pitched battle with Reggie, the show cuts back to the shot of Pete once again facing Frank—and losing—the first arm wrestling battle. The entire show, from Pete's point of view, was a flashforward—and a false one at that—and with a flash*back* inserted within it.

Furthermore, memory of media is also discussed and represented in these temporally complex television programs. For example, in *Doctor Who*, characters have constantly faced enemy attacks in the guise of mediation. In "The Time of Angels," Amy literally has to stare at a television screen in order to survive the Weeping Angel attack (this reflexively works on the audience as well, spurring us to stare intently at the screen). The Tenth Doctor travels to the early days of the television in "The Idiot's Lantern." The Jagrafess attempt to take over the minds of humanity via reality television and entertainment/news in "The Long Game." Rose is obsessed with her mobile phone. Sardick watches his past self interact with the Doctor through a home projector. The depiction of media in *Doctor Who* focuses narrative attention on the same things viewers are focusing on: television, mobile phones, and popular entertainment.

This recent depiction of complexity in temporal mechanics appears at the same moment as the rise of transgenic media in our everyday lives. We are entering a new era, where we become more reliant on technology, and we will begin to find our cultural identity stemming more and more from that technology, and less and less from the content that technology presents. Often, we find ourselves discussing not just a television program, but also the medium of delivery—web content that accompanies the show, the webisodes that bolster the narrative, the extra-textual inserts on DVD box sets, or fan-zines. Ultimately, the repeated use of and representation of memory in television content provide a useful heuristic for understanding the way meaning is promulgated across different media technologies in the digital era. We use our media to record our past and shape our future. Yet, at the same time, our memories themselves become shaped by the media that we use and view. As Sconce points out, "TV is a world that is always there, at least in unrealized form, even when the set is turned off."[53] We revisit our old media in our minds, constructed as memories within the self. Our use of transgenic media has complicated this notion of mediated memory.

Memory is a form of temporal displacement for the viewer, and on television this becomes represented through the way we view our own history. While memory itself is represented onscreen in many of these television programs, what becomes important to recognize is that it is a special kind of memory: an unreliable, fake or changeable one. With the image of an alterable mediated memory, viewers' own unreliable digital memories become more meaningful. The very human trait of memory can triumph over the *mediated* memory observed onscreen. The unreliability of media is countered by an assurance, however false, of the reliability of biology.

Notes

1 This type of mental time travel is discussed in Anne Draoulec and Marie-Paule Péry-Woodley, "Time Travel in Text: Temporal Framing in Narratives and Non-Narratives," in *Determination of Information and Tenor in Texts, Proceedings of Multidisciplinary Approaches to Discourse*, ed. L. Lagerwerf, W. Spooren & L. Degand (Amsterdam: Stichting Neerlandistiek & Münster, Nodus Publikationen, 2003): http://w3.erss.univtlse2.fr/textes/pagespersos/pery/articles/MAD03.pdf (accessed 01 Apr 2011).

2 Jameson, 70; see also Amy Holdsworth, "Television Resurrections: Television and Memory," *Cinema Journal* 47, no. 3 (2008): 142.

3 Raymond Williams, *Keywords: A Vocabulary of Culture and Society* (New York: Oxford University Press, 1976).

4 See Giuliana Bruno, "Ramble City: Postmodernism and *Blade Runner*," *October* 41 (1987), for a discussion of how memory functions to place us in history, and Scott Bukatman, *Terminal Identity* (Durham, NC: Duke University Press, 1993), 248, for memory as "constitutive of the self."

5 Mittell, "Previously On," ¶25.

6 Erving Goffman, *The Presentation of Self in Everyday Life* (New York: Doubleday, 1959), 252–56.

7 John Palfrey and Urs Gasser, *Born Digital: Understanding the First Generation of Digital Natives* (New York: Basic Books, 2008); Jonathan Zittrain, *The Future of the Internet and How to Stop It* (New Haven, CT: Yale University Press, 2008).

8 Janna Wortham, "Friendster to Erase Early Posts and Old Photos," *New York Times*, 27 Apr 2011, http://www.nytimes.com/2011/04/27/technology/27friendster.html (accessed 27 Apr 2011), ¶7.

9 Internet World Stats: Usage and Population Statistics, "Internet Users in the World," Mar 2011, http://www.internetworldstats.com/stats.htm (accessed 16 Aug 2011); Aaron Smith, "Why Americans Use Social Media," *Pew Internet and American Life Project*, 15 Nov 2011, http://www.pewinternet.org/Reports/2011/Why-Americans-Use-Social-Media/Main-report.aspx (accessed 02 Dec 2011).

10 Michel Foucault, "Technologies of the Self" (1982), in *Ethics: Subjectivity and Truth,* Vol 1, ed. Paul Rabinow, trans. Robert Hurley and Others (New York: The New Press, 1997); also Michel Foucault, "Self-Writing" (1983), in *Ethics: Subjectivity and Truth,* Vol 1, ed. Paul Rabinow, trans. Robert Hurley and Others (New York: The New Press, 1997).

11 Foucault, "Technologies," 224.

12 Ibid., 252; Foucault, "Self-Writing."

13 Douglas E. Cowan, *Sacred Space: The Quest for Transcendence in Science Fiction Film and Television* (Waco, TX: Baylor University Press, 2010), 59.

14 See Vivian Sobchack, *Screening Space: The American Science Fiction Film* (New Brunswick, NJ: Rutgers University Press, 1987), 272.

15 See, particularly, José van Dijk, *Mediated Memories in the Digital Age* (Stanford, CA: Stanford University Press, 2007).

16 Plato, *Phaedrus*, trans. W. C. Helmbold and W. G. Rabinowitz (Indianapolis: Library of Liberal Arts-Bobbs, 1956), 68–70.

17 See David Gunkel, *Thinking Otherwise* (West Lafayette, IN: Purdue University Press, 2009), 96–8.

18 Ivan Illich and Barry Sanders, *ABC: The Alphabetization of the Popular Mind* (San Francisco: North Point Press, 1988), 54.

19 Walter Ong, *Orality and Literacy* (London: Routledge, 2002).

20 Kathleen Welch, *Electric Rhetoric: Classical Rhetoric, Oralism, and a New Literacy* (Cambridge, MA: The MIT Press, 1999).

21 Sherry Turkle, *Alone Together* (New York: Basic Books, 2011), 300; see also Bradley J. Rhodes, "The Wearable Remembrance Agent: A System for Augmented Memory," *Personal Technologies* 1 (1997): 218–24; Elise van den Hoven and Berry Eggen, "Informing Augmented Memory System Design through Autobiographical Memory Theory," *Personal and Ubiquitous Computing* 12 (2008): 433–43; Thad Starner, Steve Mann, Bradley Rhodes, Jeffrey Levine, Jennifer Healey, Dana Kirsch, Roz Picard, and Alex Pentland, "Augmented Reality Through Wearable Computing," *Presence: Teleoperators and Virtual Environments* 6, no. 4 (1997): http://www.cc.gatech.edu/fac/Thad.Starner/p/journal/augmented-reality-through-wearable-computing.pdf (accessed 14 Mar 2011).

22 Turkle, *Alone Together,* 300.

23 Ibid.

24 Ibid.

25 Vannevar Bush, "As We May Think," *Atlantic Monthly,* July 1945, http://www.theatlantic.com/magazine/print/1969/12/as-we-may-think/3881/ (accessed 15 Mar 2011), ¶6.

26 Ibid., ¶54.

27 Turkle, *Alone Together,* 302.

28 Cliff Atkinson, *The Backchannel: How Audiences Are Using Twitter and Social Media and Changing Presentations Forever* (Berkeley, CA: New Riders, 2010), 128.

29 Ibid., 128.

30 Ryan, 124–25.

31 http://911digitalarchive.org/.

32 Ekaterina Haskins, "Between Archive and Participation: Public Memory in a Digital Age," *Rhetoric Society Quarterly* 37 (2007): 401–02.

33 Michael Strangelove, *Watching YouTube: Extraordinary Videos by Ordinary People* (Toronto, CA: University of Toronto Press, 2010), 33.

34 Ibid., 138.

35 Ibid., 39.

36 Mittell, "Previously On," ¶3.

37 Ibid., ¶7.

38 Rick Altman, *A Theory of Narrative* (New York: Columbia University Press, 2008), 292.

39 Espenson, 46.

40 Bordwell, 91.

41 Umberto Eco, "On the Ontology of Fictional Characters: A Semiotic Approach," *Sign Systems Studies* 37, no. 1/2 (2009): 97.

42 Mittell, "Narrative Complexity," 36.

43 Mieke Bal, *On Story-Telling: Essays in Narratology* (Sonoma, CA: Polebridge Press, 1991), 78.

44 See Wayne Booth, *The Rhetoric of Fiction* (Chicago, IL: University of Chicago Press, 1961).

45 Max Shulman, *Max Shulman's Large Economy Size* (New York: Doubleday & Company, 1948).

46 A clip show uses largely recycled "best-of" clips from the history of a television show, and is often presented during an anniversary of a television series or as a "special."

47 Created memories are not unique to the first decade of the twenty-first century, however. For example, in the *Star Trek: TNG* episode "The Inner Light," Captain Picard is implanted with *real* memories which have been created by a culture known as the Kataan. These memories record the final decades of an advanced civilization as seen through the guise of Kamin. Picard, who in these memories lives several decades as Kamin, wakes up on the bridge of the Enterprise to find that only 25 minutes have passed during his lifetime as Kamin. Yet, he retains those memories of his *other* lifetime, symbolized in his "newfound" and practiced ability to play the flute.

48 Karen Hellekson, "Memory and the 1996 American Remake of *Doctor Who*," in *American Remakes of British Television: Transformations and Mistranslations*, ed. Carlen Lavigne and Heather Marcovitch (Lanham, MD: Lexington Books, 2011), argues that memory is also key to the 1996 US/British co-produced *Doctor Who* TV movie.

49 See "Desmond's photograph," *Lostpedia*, http://lostpedia.wikia.com/wiki/Desmond's_photograph (accessed 01 Feb 2011).

50 Bordwell, 79.

51 Paul Nahin, *Time Machines: Time Travel in Physics, Metaphysics, and Science Fiction* (New York: Springer, 1993), 129.

52 Roland Barthes, *S/Z*, trans. Richard Miller (New York: Farrar, Straus and Girous, 1974).

53 Sconce, *Haunted Media*, 174.

Three

Personal History as Explorational Media

> *In other words, a character within a story, or even a narrator, has, in general, no way of knowing whether or not he is in the past tense narration of a story, or is instead in the present tense (or some other tensed state of affairs) and merely reflecting on the past.*
>
> —Yu, How to Live Safely in a Science Fictional Universe, 33

For Yu, as both an author and as a character, the titular "Science Fictional Universe" of his book becomes problematically populated with personalities. The simplicity and accessibility of time machines in this novel creates a population of self-explorational visitors. Since they can only travel in their own lives, the characters become self-obsessed, narcissistic, and egotistical. As uni-linear time travelers, we too are all characters in our own life stories, but we can only travel with an inevitable forward momentum. We do narcissistically back travel in time, however, when we use our memory or our transgenic media to examine our past selves online. As I will show in this chapter, contemporary complex television reflects this narcissism using characters who travel in time and become a part of their own past lives. This type of televisual representation of our culture's tenuous relationship between the *self* and the *mediated self*, what I'm terming *explorational media*, reflects the schism between who we are and who we present ourselves to be through our transgenic media use. In this chapter I examine contemporary television as a complement to and a heuristic for this personal narrative format. And in doing so, I examine the very complex notion of "character" as it relates to viewers of temporally complex television. The personal narrative has become a powerful form in our transgenic media—the ability to broadcast the self has never been more widespread than it is today. Yet, these personal narratives also present troubling glimpses into our mutable history. Yu asks an important moral question with which few novels of time travel engage—namely, what happens to one's sense of self if that self is constantly in temporal flux? Who are you, if you are "outside" time? Explorational media ask us to consider similar questions about the relationship between our self and representations of our self.

In the last chapter I looked at how characters' memories are represented through the temporal displacement of narrative information in complex television. In this chapter I hone in on the way temporally displaced narratives personalize "memory" as an aspect of character temporality. In this way, I explore a second dialectic of transgenic media, between the *personalization* of self and the *publication* of the self.

As both a literary and a televisual topic, the notion of character has been little explored in scholarship.[1] Indeed, in any definition of narrative, "virtually no attention has been paid to the notion of character."[2] As Phelan describes, characters are rarely examined in their own right, but rather become subsumed as aspects of narrative analysis or genre analysis.[3] Altman notes the crucial importance of character for any narrative assessment.[4] And, in fact, it's difficult to translate any discussion of character in a literary analysis to a televisual one, as literary explorations tend to focus on the role of the narrator in the construction of character, a role that Kozloff argues is often an "intangible" presence on television.[5] The previous literature that examines characters on television has tended to analyze them

structurally, using a typography of particular character traits.[6] This is not uncommon, for television scholars have often used a form of literary criticism as the basis for their analysis.[7] For example, Pearson identifies six traits scholars can use to help illustrate character uniqueness and audience identification, looking closely at one character from *CSI* (2000).[8] In her 2009 contribution to *Reading Lost*, a critical analysis of the eponymous television series, she does a similar reading of John Locke.[9]

Indeed, while it may be difficult to separate out the character threads from the narrative tapestry in novels and short stories, it is even more difficult to do so in contemporary narrative television. Television is a character's medium. Even if characters become more standardized across shows, conforming to convenient stereotypes and market appeals, they still form the basis for audience identification. After all, if we didn't like the characters, we wouldn't watch the program (no matter how avant-garde or temporally displaced the narrative). We grow to care about characters and, from this emotional bond, become attached to televisual programming.[10] Part of our attachment comes through the repetitious nature of television, both through syndication as well as through rerun scheduling. The serialized nature of television character garners viewer identification through repetition and familiarity—the more we see, the closer we feel. We're used to seeing characters every week, and they become staples in our lives through an "idea of continuity."[11] But attachment can come through the *style* of episodic television content as well. Characters often act in surprising or unexpected ways—and our attachment to them reveals itself when we see them "acting out of character." For Chatman, such an "open theory of character ... treat[s] characters as autonomous beings, not as mere plot functions."[12]

Some good scholarship has already been done on the complexity inherent in character on television. For example, Allrath explored the idea of character complexity in her chapter about alternate worlds and doubling in *Buffy the Vampire Slayer* and *Angel*, arguing that "the traditional unity and stability of character is subverted in a number of ways" in the shows.[13] Characters are often doubled, sometimes in the same body (Angel/Angelus) and sometimes from alternate universes (Willow/Evil Willow), and some characters have dual roles (Giles is both Watcher and Ripper). Ultimately, she concludes that "the notion that characters are unique and stable individuals ...is radically subverted" by creator Joss Whedon.[14] Given Whedon's propensity for complex and nuanced characterizations, this comes as no surprise; and is an argument I will revisit when examining his *Dollhouse*.

In this chapter I focus on the way that complex television characters reconcile their own past, and compare this to the way contemporary transgenic media users reflect upon and change the representation of their own historical existence. In doing so, we learn how to engage with our past in a meaningful way.

Personal Narrative in Transgenic Media

Although some critics might call it narcissism, today's transgenic media allows users to write personal, introspective and private thoughts in public arenas. For many the "diary-writing" of blogs, the personal videos on YouTube and the self-centered communities on Facebook offer a unique glimpse into people's own private lives—a view that until a few years ago, would have been much different. What has changed is not the desire to leave traces of the self across media, but the twinned roles that privacy and publicity have in shaping the format of that self. People have always been introspective, have always kept diaries, and have always stored images of themselves and their loved ones. The technology may change, but the impetus to *record the self* remains. When diaries become public, of course they might seem self-indulgent. But as Shirky describes, the reason so much conversation over transgenic media sounds silly—Twitter updates about last night's dinner, MySpace statuses that discuss clothing, YouTube comments that reference inside jokes—is that the writers are not talking to you—they're talking to each other.[15] It's like overhearing half a conversation between two friends: the possibility that you'd find their idle chitchat stimulating is highly unlikely.

There has been a shift in conversation and introspection—the two main forms of writing, according to Foucault—from keeping them private to making them public. What has that done to our senses of *self*, our "offline" identities?[16] Take, for example, the autobiographical technology of the web. Given a basic knowledge of transgenic technology, one can open, in quite a short period of time, a public blog, a public social networking profile, a Twitter account, and a YouTube account. Throw in an LTE cell phone with recording capabilities, and there isn't a moment of one's life that can't be uploaded. A user could create his or her own MyLifeBits. None of this material, however, disappears naturally. Even if one deletes it from his/her site, once it's public, it's public permanently. Forget Las Vegas—what happens on Facebook, stays on Facebook...unless someone copy/pastes it to Wordpress, LiveJournal, Tumblr, YouTube, Vimeo, Twitter....

Our offline identities are changing because of this increase in publicity and self-publication online. Anytime I want to, I can go onto YouTube and watch videos made fifteen years ago starring me. That's a public record of my life—yes, others can see it, and yes, it may make me embarrassed to view my Junior High School production of *Romeo and Juliet*, but importantly *I can always view it*. And through this constant access to my personal history, the web doesn't just become a repository of information and useful factoids. It becomes a space of my own: a house for my history. I can use the web anytime to see who I was, and to explore whom I could become.

Our contemporary television programming reflects this self-introspection engendered by the web. By using temporal displacement, complex television narratives can show audiences a character's memories, flashing back to previous points in the character's life, or can demonstrate introspection in a more visceral way, through a character's time travelling. This introspection helps create a sense of autobiography for the characters in a complex narrative. I want to look now at the way "character" itself is constructed on television as a surrogate for the viewer. When television programming focuses on characters who travel back into their own lives, it becomes a means for a characters' self-reflection. And just as transgenic media allow for the permanent introspection of the user, so too does television allow for this narcissistic meditation for the character to develop.

Character on Television

Glen Creeber shows that "the television serial (with its long-form, episodic nature) arguably [possesses] a greater potential" for character studies, and "long-form drama is intrinsically better suited to explore and dramatize the complexity of character psychology as a whole."[17] As a component of narrative theory, "character" is crucial for eliciting audience identification.[18] Indeed, "for entertainment to be able to exist, a basic need must be satisfied: the need to come into contact with or relate to media characters."[19] As Jonathan Cohen notes, identification is a key moment in a narrative text, whether literary, cinematic, or televisual: "*identification* is a mechanism though which audience members experience reception and interpretation of the text from the inside, as if the events were happening to them."[20] Through identification, we can more fully reflect upon the immediate and personal impact of media messages. This is not a psychological malady.[21] Rather, it is a method for understanding the complex relationship between viewers and media creators. Cohen goes on to link identification with physical manifestations: while identifying with a character, "an audience member imagines him- or herself being that character and *replaces* his or her personal identity and role as audience member with the identity and role of the character within the text."[22] While I disagree that an actual *replacement* occurs, I do think that there is a form of identity *augmentation* between character and viewer. The knowledge that the characters gain augments the audience member's knowledge. This mutual experience bonds the character to the viewer. As literary scholar Blakey Vermeule notes, literary characters "are the greatest practical-reasoning schemes ever invented. We use them to sort out basic moral problems or to practice new emotional situations. We use them to cut through masses of ambient cultural information."[23] The best characters, the ones we remember throughout our life:

> may cause us to occupy their world a bit longer. And while we linger—reflecting on their traits, mulling their relationships, gauging their development, or even weighing their goals—we find ourselves reading more deeply. The characters we pause to consider can guide us through their stories, helping us to understand plots and ponder themes; ...characters can make us better comprehenders.[24]

It's not an accident that a variety of characters exist with whom many people feel close ties. Media producers often "create characters with whom the audience are meant to interact to enjoy books, films, or television programs."[25] Perhaps this is because, as Goffman describes, an individual is a performer as well as character in his or her own right, "whose spirit, strength, and other sterling qualities the performance was designed to evoke."[26] We see ourselves as characters in our own lives. Television scholar Horace Newcomb notes that television characters invoke a "sort of intimacy, ... the possibility for a much stronger sense of audience involvement, [and] a sense of becoming a part of the lives and actions of the characters they see."[27] Characters may influence audiences, but audiences imbue characters with life.

As Bal points out, audiences "'see' characters, feel with them and like or dislike them."[28] Additionally, character becomes a form of "play" that helps socialize readers into more complex social interactions.[29] Readers learn how to act in various situations by using characters as a form of "avatarism" to make life more comprehendible.[30] Ira Newman echoes this assumption, calling characters "virtual people" who elicit puzzle-solving behaviors from readers.[31] For Marie-Laure Ryan, the notion of "character" assumes a more emotionally charged level of behavior for the reader: "Characters possess among other aspects the mimetic dimension of pseudo-human beings."[32]

Ryan's use of the term "mimetic" refers here to James Phelan's delineation of the three roles that characters can have in a narrative:

mimetic, synthetic and thematic. A mimetic understanding of character is one that represents aspects of a "real" person—that is, a character's traits mirror those of "real people," and the author asks the reader to reflect on the realism of the character. The mimetic dimension of character reveals an audience's identification with the character.[33] Eco takes this one step further, writing that when "we certainly identify ourselves with the cases of fictional characters," we necessarily also need to "forget" that it was created by an author:[34]

> Our problem is why we can be emotionally moved by the deeds of fictional characters. In spite of every logician nobody is supposed to weep because *Tolstoj wrote that Anna Karenina died.* This is none of our business. One feels moved, at most, because *Anna Karenina died*—even if one ignores that it was Tolstoj who first wrote it.[35]

A synthetic understanding of character, conversely, relies on the audience's understanding that the character is a created, artificial construct. For Phelan, the power of the synthetic lies in the way the character foregrounds the importance of art and aesthetics in narrative form.[36] We can admire the fictionality of a character as a way of appreciating the work that went into its construction. For example, we may value the performance of Alex Kingston as *Doctor Who*'s River Song, or the way Steven Moffat writes her time-twisting plotline. The synthetic and the mimetic functions are not opposites—we can admire artistry while also being emotionally moved—and this is how fictional characters become, at least for Eco, semiotic objects.[37] Roser et al. show that "readers understand that characters cannot be lifted away from the narratives that enfold them. We know that settings and plot shape and bind characters to contexts that have a great deal to do with their development."[38]

Finally, a thematic understanding of character focuses on the larger symbolic representations a character can take. For instance, a character might be taken as a representative of a certain class of people—Hurley on *Lost* often represents the audience, asking the questions the audience wants to ask. Peter on *Fringe* fulfills the same role, explaining Walter's techno-babble to a lay audience.

Thematically, Hurley and Peter have roles that go beyond their mimetic and synthetic qualities.[39] It's important to recognize that "the actions and events of a character's fictional life provide data through which we can imagine who the person is—in the sense of recognizing his or her psychological or moral character," and this recognition leads us to learn from them.[40]

The mimetic, the synthetic, and the thematic understanding of character help shape the way characters have been interpreted, especially in literary criticism. Yet, at their base, all three character roles depend on the author's creation of character, and argue that the author of the narrative has given that character meaning. Indeed, for Phelan, the role of the character emerges mainly through authorial intention: as he writes, "when an author creates a character, [he/]she creates a potential for that character to participate in the signification of the work through the development of the character."[41] Yet, there is a more complex interaction between character, author, and audience at play in televisual narratives as opposed to the literary standpoint of Phelan.[42] One issue with this auteurist understanding of character is that, as Nikolajeva states, audiences "can easily ascribe to [characters] features that the author had no intention of providing."[43] Indeed, it is the audience's prerogative to "re-read" characters with different meanings; to discover new readings; to provide alternate interpretations of actions. In televisual media, the roles of characters have become more complicated than what Phelan describes. Newman shows that the idea of "character entails transworld applicability," as "the same or a similar character can be exhibited through the frameworks presented by a number of different worlds of events."[44] Newman's concept of the "transworld" television framework highlights the difficulty with tracing character across transmedia as well, which entails a different set of mechanisms than what Phelan designates.

Specifically, by integrating Seymour Chatman's notion of an "open theory of character" with Phelan's delineation of character roles in literary texts, a fourth type of character role, one specifically designed to reflect televisual texts, becomes viable. This fourth role, what I'm terming the *metaphoric,* describes how a character *only* achieves meaning through audience identification. This is different from the thematic understanding because it relies inherently on a reader-centric model of characterization. By viewing characters as autonomous beings, and by highlighting the role of the audience in this construction, a metaphoric character underscores identification. That is, a character may have an explicit role to play in the construction of the narrative story, but it is only through the reconstruction "by the audience from evidence announced or implicit in an original construction and communicated by the discourse" that the open character becomes realized.[45] Indeed, as Currie puts it, an external perspective is "essential" to coming to any sort of understanding of the aspects of narrative.[46]

Metaphoric characters have been a part of television discourse since the early days of the medium. For example, one such metaphoric character is the Doctor from *Doctor Who*. Early in the series' history, the producers wanted to change the lead actor on the show but keep the character the same. Thus, they introduced the idea that the Doctor could "renew," or alter his appearance and personality while remaining essentially the same character. Given the nature of regeneration (as it canonically became called) as a key component of the series, the character of the Doctor has always been in flux. Any sense of character stability only comes through the interaction of the audience. The producers have to rely on an active and engaged audience to understand the switch of the main actor every few years. The Doctor must inherently be open-ended: his personality shifts and changes with every regeneration, and thus he becomes new even though the character is very old. Currently on his 11th body, the Doctor's regeneration has allowed new generations of fans to experience the "same" character across five decades of media texts.

Crucially, the narrative of *Doctor Who* is also altered by the sense of regeneration as multiple Doctors can meet, for example, in such special episodes as "The Three Doctors," "The Five Doctors," and "The Two Doctors." Most recently, the 2007 Children in Need charity special "Time Crash" highlighted the temporal displacement of these character reunions. The Tenth Doctor (David Tennant) crashes his ship, the TARDIS, into the Fifth Doctor's (Peter Davison) TARDIS. The resulting time crash ruptures the space/time continuum and both versions of the same character meet face-to-face. The Tenth Doctor, of course, knows who the Fifth is, but the Fifth had no idea about the Tenth (he claims the Tenth Doctor must be an obsessive fan). The solution to break the two ships apart comes directly as a result of this temporal displacement. The Tenth Doctor, aware that this event must have happened in his past, "remembers" how he watched himself (Fifth watching the Tenth) and rights the ship in front of the Fifth (who then remembers it for his future).

Importantly, the nature of the Doctor's complicated relationship with himself only functions through the interaction with an audience already familiar with the program's stability. "Time Crash" directly bridges the divide between the Classic *Who* (1963–1989 series) and the updated New *Who* series (2005–). The two eras do bridge occasionally, as in the New *Who* two-part episode "Human Nature/The Family of Blood," when the Doctor draws a picture of his former incarnations in his diary. But to make sense of that drawing, or of the presence of the Fifth Doctor with the Tenth, there must be a form of identification between the audience and the character. Metaphoric characters necessitate audience identification, but that meaning only becomes realized through the interaction of the audience and the text.

Character Temporality

The relationship of the Doctors in *Doctor Who*, especially in crossover episodes like "Time Crash," represents a further type of temporal displacement in contemporary television: "character temporality." This form of temporal displacement features characters who exist separated from their own temporal context, estranged from their "normal" time-space arena. In shows with this form of displacement, characters can be disembodied, their minds transporting them imaginatively to a different time. Or, they can be taken bodily from their own temporal realm and transported else*when*. Shows with displaced character temporality tend to aggregate temporality and spatiality into a time-space amalgam, in which a character's temporal displacement is mirrored by a similar spatial distortion.

The temporal displacement of complex characters in contemporary television content hinges on the metaphoric role of character within narrative. An increase in audience identification via transgenic media highlights the complex relationship between *user*, *character*, and *avatar* as a unique mashup. In one of the classic texts about the relationship between narrative and the virtual world of cyberspace, Murray details how the digital medium produces a metaphoric role for electronic characters.[47] Digital characters are ascribed with traits by their users, adapting to the interactions provided by those at the keyboard. The avatar, a digital creation that has its own identity but is also based on another's identity, becomes a façade externally played with and internally experienced.[48]

We live in an epoch of identity dislocation, where our own identities (partly through this complex avatarism) are in flux. Although transgenic media may be the most prevalent means of exploring the self today, people have used traditional media to examine who they are for centuries. Foucault reminds us that a simple pen and paper (or quill and papyrus) were enough to explore issues of the self.[49] And as Goffman describes, explorational media don't just explicate the self for others' edification, but also scribe the self for our own personal growth.[50] That is, we write down our social characteristics and our interests on Facebook not only to connect with others, but also to reflect an immediate and personal representation of ourselves to our self. Yet, any identity performance is always limited by the social context in which it exists: Goffman argues that any identity performance constantly tests these limits through "concealed practices" or "particular performances."[51] Indeed, postmodern theorists would say that we've lost any sense of a coherent identity.[52] As Manuel Castells has shown, the rewritable functions of the Internet allow people "who are in the process of discovering their identity, of experimenting with it, of finding out who they really are or would like to be" to understand "identity construction and experimentation."[53] Scott Bukatman calls this a sense of "terminal identity," where "we find both the end of the subject and a new subjectivity constructed at the computer station or television screen."[54] In other words, when we lose sight of who we *are*, we regain it by externally experiencing it mediated back at us from the screens in our lives. We feel overwhelmed by events that are out of our control: the presence of catastrophic illness, environmental disasters that seem too big to clean up, the death of family members. It's easy to give up caring when the problems are so monumental. Bukatman's "terminal identity" positions the subject in contemporary society at the nexus of mediation and identification. By defining our reality through a mediation, "a new subject has emerged: one constituted by electronic technologies, [and] also by the machineries of the text. Terminal identity is a transitional state produced at the intersection of technology and narration."[55]

As part and parcel of this terminal identity, individuals in society find themselves displaced from a sense of relevant historical context. Vivian Sobchack examines this notion of temporality in the final chapter of her *Screening Space*: she argues that there has been a "collapse of ... temporal relationships that formulated time as a continuous and unifying flow" in our contemporary postmodern culture. Riffing on Jameson's discussion of the "weakening of historicity" in postmodern life, she argues that shifts in conceptions of temporality have "several existential and aesthetic consequences," not least of which is a collapse of the "coherence of personal identity, history, and narrative," a consequence we can see reflected in much of our contemporary television entertainment.[56]

Explorational Media

The break in temporality seen by these postmodern theorists manifests in characters on contemporary television who experience life out-of-order by travelling in time. Two exemplar characters come from the BBC television shows *Life on Mars* and its sequel *Ashes to Ashes*. Both shows demonstrate how individuals can re-examine their own lives through personal narratives. Tyler and Drake also serve as useful avatars for the viewer, in that the exploration each pursues matches a similar personal history exploration with which users of transgenic media engage. In both shows, Tyler and Drake encounter their own selves as children, face their younger parents, and revisit childhood events. In this way, their memory, their personal narrative, must be reformed as they literally face themselves and change their memories as they are formed/forming.[57] These characters self-reflexively reappraise their past and restructure their present, and their personal history becomes a continually updated narrative. Ultimately, both protagonists fail at changing their own personal narratives, as the childhood traumas they experience/d continue to haunt them through adulthood. They are forced not to change their story, but to augment their memories of it, a realization that both foretells and reflects transgenic media's insubstantiality. Not only do the plots of these two shows represent the nostalgia of the baby boomer audience at which they are targeted, but they also encompass television's use of transgenic media characteristics.

In *Life on Mars*, Manchester police officer Sam Tyler is hit by a car in 2006 and awakens in the year 1973. He finds himself working for the same police department thirty-three years in the past, under the command of Chief Inspector Gene Hunt. The rationale behind Tyler's exile to the past is the resounding serial arc of the show, and we can observe much character complexity within Tyler's temporal displacement. At times Tyler—and thus, the audience who identifies with this metaphoric character—doesn't know whether he has really time travelled or if he is lying in a hospital in a coma, dreaming of this time period. To investigate, Tyler starts to venture into his own life when he was a child growing up in Manchester in 1973. A crucial scene in the finale of the first season finds Tyler talking to himself as a child: and during the conversation, he realizes that his memories are in fact being formed at that moment as he communicates with himself. The temporal loop closes as past, present, and future Sam Tyler all converge on this one temporal location: a precise demonstration of Jameson's simultaneous "presents in time."[58]

In *Ashes to Ashes*, the sequel to *Life on Mars*, Alex Drake, the police psychologist hired to work with Tyler in 2006, is shot and wakes up in 1981—during the time period when she was just a child. In this show, however, the temporal complexity increases, as the first series' plot revolves around the death of Drake's parents. Images of the elder Drake's memories from that time period are interspersed with representations of the young Drake's experiences. The audience sees that those images are changing due to the elder Drake's presence back in that earlier time. In other words, Drake comes to terms with her parent's death by visiting the site of their killing in the past and then *revisiting* it in her memory while there. Saliently personified by the character of Gene Hunt, Drake finds that memories of her guardian are replaced by memories of Hunt—and the audience can observe how both sets of memories form simultaneously.

In the first series of *Ashes to Ashes*, Drake (living in 1981) befriends her own mother without revealing their "true" relationship. In doing so, she is given the opportunity to explore her childhood home from the vantage point of an adult: she sees her old tape player, her posters of pop stars, and even reads a portion of her own diary. The play with temporality in *Ashes to Ashes* allows Drake a chance to experience her life anew; to revise as she relives. Not many of us get a chance to explore our past with such visceral tangibility, and in portraying Drake's exploration, *Ashes to Ashes* not only plays on our feelings of nostalgia, but also harnesses our collective experience of exploring our own personal pasts.

To identify with this form of character traditionally, according to Eco, we either have to see the fictional world as a daydream, positing specific characteristics as aspects of an auteuristic subconscious, or we must in some way behave as if we were one of its characters.[59] Tanya Modleski describes a third type of identification, one based on seeing the characters as intimates or friends instead of as surrogates.[60] Modleski's view of character identification parallels a type of parasocial interaction with the characters, where viewers take the representation as the reality and "form bonds with the characters…[which] mirror the social bonds between people that meet in the 'real world."[61]

Yet, integrating the metaphoric role of character with the sense of personal exploration that character temporal displacement engenders reveals a deeper connection between character and audience. In effect, by relying on a sense of nostalgia, these shows illustrate a journey on which many audience members may embark. By exploring their pasts, Tyler and Drake both critically re-examine their presents; similarly, by exploring our pasts through transgenic media, we are able to re-examine our roles in contemporary society. The appeal of shows like *Life on Mars* and *Ashes to Ashes* lies not just in enjoying the taut plotline or witty dialogue; it lies in re-experiencing a time period that exists in one's memory. For many viewers of these shows, myself included, part of the joy of the show lies in the nostalgic trip to the past: I'm enthralled not just in the period costumes or typical (lack of) technology, but also in the stylistics (the credit sequence to *Ashes to Ashes*, for example, recalls many an 80s action/adventure program from my childhood, complete with electric guitar riffs and videographic titles).[62] Relying on temporal displacements to enhance viewer self-exploration helps solidify the connection between audience and character. As David Lewis explains, temporal discontinuity lies at the heart of any narrative experience. He writes, "if [you] should change your mind about the possibility of time travel, there will be a difference of opinion between two different temporal parts of you, the stage that started reading and the subsequent stage that finishes."[63] Limiting temporality to the *character* of a time travel plot ignores the important effects of temporal distortion on the *viewer* as well.

There has always been a desire in our culture to record our history and share it with others, but it has never been as easy or as public as it is now. At the same time that *Life on Mars* was produced, the Pew Research Center found that the practice of blogging rose among older adults and social networking also rose significantly across all age groups.[64] Through this mutual swapping of information, we form a more thorough and communal understanding of our present—the friend that sees my post about the past, then comments on the connection she shares with me. Our own pasts have become temporally displaced through transgenic media, as we are able to retell the events of our life in (nearly) real-time. I can post a status update from my smart phone the very instant I purchase a new shirt or experience a delicious sandwich. Imagine the following scenario: My friend can use Foursquare to "check-in" at a coffeeshop, a service recorded for posterity. Another friend checks-in at the same coffeeshop. A few weeks later, they both change their statuses on Facebook to "in a relationship." A few weeks after that, photos of a vacation together are posted, tagged, and grouped on Flickr. They synch their Goodreads accounts. Their public Amazon.com wishlists register larger bedsheets, more Tupperware, and some cheap furniture. A video tour of their new, empty apartment shows up on YouTube. Then, on Picasa a photograph of a ring appears. Many years down the road, a mother can post images of her ultrasound, which the father can "like." These immediate re-tellings of important events from people's lives form a narrativized account of who these people are. Similarly, Creeber shows that many long-form narratives pit "social reality itself ... against subjective, individual and multiple perspectives."[65] Today's long-form narratives, in other words, "more accurately reflect" our role in the world's "social inconsistencies and moral uncertainties."[66] But it's not just our own narratives that shape our experiences—it's what we see on screen as well. As Sandvoss shows, in many ways we identify our sense of self with mediated narratives. Changes in mediated narrative structure illustrate changes in our conceptions of ourselves.[67]

Metaphoric Dimensionality

The revision of memories in temporally displaced television narratives becomes a dual-temporal device when the depiction of one time frame appears alongside the depiction of another. This representation of memory becomes emblematic of how technology can be used to "re-present" memories. For example, we can alter and change digital photographs, literally changing our representation of history. As he points out in *Camera Lucida,* Barthes may feel a connection to a photograph of his mother, but would he have digitally colored her hair? On television, the use of the viewer's long-term memory is made even more visceral, as successful shows engender nostalgia in long-time viewers. Rare is the show that reaches its 100th episode without some form of clip show retrospective. As Holdsworth writes, "the popularity of nostalgia programming (i.e., the clip show or list TV [e.g., *The 100 Best Moments of TV Comedy*]) and the scheduling of retrospective seasons...the phenomenal growth of the TV-DVD market, and the development of online television archives" marks "the currency of television memory and nostalgia."[68] Television programming which relies on memory, like *How I Met Your Mother,* features an air of nostalgia. One could interpret the show's extensive flashback structure as a nostalgic look at the first decade of the 21st century. The nostalgia of Future Ted is often contrasted with his own self-censorship, as he is speaking to his children (the diegetic recipients of the memory). So, instead of talking about "smoking pot" in college, he refers to "eating submarine sandwiches." To illustrate this nostalgic self-censorship, the characters are portrayed eating—and getting stoned on—the sandwiches.

An additional type of character temporality occurs when characters physically and nonlinearly move from one temporal location to another. In contemporary television programming, there are two main ways characters enact temporal displacement. First, the character him/herself actually jumps in time, creating a link between the present moment in the narrative and a past or future moment.[69] Or second, the narrative can link characters through time via a complex network that changes the temporal framework by depicting scenes more than once, re-airing dialogue heard from a different perspective, and/or repeating scenes from more than one view point.

Character Jumps in Time

Perhaps the most common type of character displacement in time comes from its literal interpretation: moments when a character jumps time frames and is removed from one moment to be placed in another. Characters may be time travelers, as in *Life on Mars*, *Ashes to Ashes*, or *Doctor Who*, and physically move from one time frame to another. Or, characters' consciousnesses may shift temporal locations, as does Desmond's on *Lost* and the planet's population in *Flashforward*. Either way, complex narratives that focus on this type of character dislocation often follow the adventures of these "out-of-time" characters as a way of focusing attention on the differences between the present and the "new" time frame, as Nahin has shown.[70] The coherence of an audience to a temporally-displaced character functions because the character is not the only thing displaced in time: by following the character's adventure, the representation of the narrative in the show also jumps in time, and the audience likewise is forced to re-assemble the timeline in a non-linear manner. In doing so, the audience becomes privy to the same mental gymnastics encountered by the protagonist, a metaphoric character trait. A quintessential example is presented by the first episode of *Life on Mars*. The episode begins with Tyler, in 2006, attempting to catch a serial killer who kidnaps, molests, and then kills young women. In fact, this is the case that initially leads Tyler to travel in time: in the course of the investigation, the same killer kidnaps Tyler's partner Maya. Frustrated at his inability to work the system, Tyler gets out of his car on the road to vent his anger, and in doing so is knocked unconscious by a passing motorist.

Tyler quickly awakens in 1973. He is dressed in period clothes, engages in a humorous conversation with a local bobby, and learns that he has jumped back in time. Walking into the old police station, Tyler is immediately thrust into their current case: a serial killer with the same modus operandi as the one who took Maya. Using his knowledge of the case from the future, Tyler is able to solve the case in the past: the killer in 1973 was the neighbor of a young boy who grew up to emulate him. Upon being released from a maximum security mental hospital (thirty-three years later), the man kills again. Yet, in solving the case, Tyler inadvertently (or fatedly) parades the 1973 assailant past the young boy, glamorizing the case and causing the boy to become enamored of the criminal. The boy grows up to be the prime suspect in the case, who eventually wastes Tyler's (future) time as a red herring for the crime.

In a more classically themed television program, this "solution" to the time travel riddle might have catapulted Tyler back to the "correct" time, in order that he might save Maya. Yet, *Life on Mars* continues the temporal displacement across the rest of the series: Tyler not only stays in 1973, but even when it is revealed in the final episode that he has been in a coma the whole time, he chooses to commit suicide to go back to the past. Tyler's temporally-complex investigation, drawing on future memories and past techniques, presents the audience with a metaphor for its own use of transgenic media. A running theme throughout *Life on Mars* is Tyler's investigation of his own family. In the last episode of the first series, Tyler finds his father—a man absent from his early childhood—in order to convince him to come home. Yet, during his conversations with his father, he learns that the elder Tyler is a person of interest in an investigation for the murder of a bookkeeper's clerk: is it better to be on the lam from the law or reside in jail? Tyler's encounters with his father force him to revise his own memories; to make sense of the knowledge he had previously encountered in memory. Tyler tries to find comfort in his experience of the past and he explores his own memories as they develop with and for him.

Both Tyler and Alex Drake are given the chance to develop meaningful, adult relationships with their parents: a feat Rickman says is characteristic of time travel plots.[71] Similarly, we can see our parents in a new light as "friends" on Facebook, and examine their past through photographic evidence on Flickr. We flip through a Picasa account instead of travelling in time, but the impetus is similar. Transgenic media record and portray our past in a way that speaks to constant reinterpretation of the present, what the Pew Research Center describes as the constant "shroud" of our digital lives.[72] We can look at our history on MySpace and see past mistakes; or follow fragments of conversations between friends or ex-lovers and reinterpret their meanings, given this new information. Importantly, it's not that this type of interaction is unique to transgenic media, but rather that it is now much more common and available.

On shows with character complexity, another type of character dislocation exists in which characters remain affixed in time but the context around them time shifts. Time shifting can be seen as the inverse of character temporality. The character/narrative complexity of *Life on Mars* and *Ashes to Ashes* is mirrored in the show *Eureka*. Instead of finding themselves outside of time, the characters in *Eureka* find themselves in the same time, but the world around them has changed. In the fourth season of *Eureka,* five of the series regulars (Jack, Allison, Henry, Jo and Fargo) are accidentally transported to their hometown of Eureka in 1947. Upon their return to the "present," they find that slight changes in the past have had an enormous effect on the present. For example, Allison finds that her son Kevin is no longer autistic and life-long bachelor Henry discovers that he is married. Although the characters are back, their immediate context has shifted.

Such temporal leaps are not uncommon in television: we have seen time travel depicted on the screen since at least *Captain Z-Ro* in the early 1950s; and even today it's a common trope in many science-fiction shows. The pleasure of these narrative spectacles lies not just in experiencing the ongoing serial plot about what happens to the characters, but also in acknowledging and experiencing those temporal changes—the paradox of the Butterfly Effect—as moments of operational aesthetics. It is interesting that Fargo is now head of Global Dynamics, but it's only narratively and aesthetically pleasing if we also know that he *used* to be the butt of GD's workforce's jokes.

The changing context of time travel can have profound effects on the character involved. In the *Doctor Who* episode "Blink," for example, policeman Billy is sent back in time by the Weeping Angels, a race of monstrous aliens who thrive on "potential energy"—with one touch they can send a character back in time and feed on the time that the person missed experiencing. Billy, as a character sent only a few decades back in time, ends up doubling his own timeline. His trip from 2007 to 1963 finds him living through the same period of time twice. He cannot act to meet himself, as this would (as the Doctor claims), destroy two thirds of the universe. But he does exist in two forms at the same time, his context changing in his "new" time.

The duality of Billy's existence speaks to a similar duality in contemporary transgenic media: the way explorational media function on the social network profile. Two profiles of the same person can cohabitate—and, in fact, many people have two profiles in order to promote different aspects of their selves. I may have a Facebook profile for my personal life and one for my professional life. I also might have a MySpace profile, and Orkut profile, a Bebo profile and a Habbo profile. Each of these can show a different aspect of my personality, a different virtual identity.[73] Billy's dual existence in the same temporal plane speaks to this duality in our own lives today. There is a larger cultural reaction happening in both television and transgenic media that reflects the issue of doubling of identities and personas.[74] By viewing our profiles, we see the dynamic between the self and the mediated self, or the way that we present different aspects of our personalities online.

Doctor Who relates to *Life on Mars* and *Ashes to Ashes* not just because they are all BBC shows, but also because they all deal with memory as constituent of personality. The fifth series of *Doctor Who* also illustrates the importance of character when the Doctor must travel through his own memories in order to remind companion Amy Pond of his own existence. In the finale of the 2010 series, the Doctor has been privy to and blamed for the destruction of all of history: an event which is troublesome in its complexity (I will describe this in more detail in Chapter Five). The only way to reestablish a link between the events as they happened in time and the *knowledge* of those events (that is, the way those events become actualized by the characters), is if the Doctor can travel back through his own memories of Amy, on whom the universe has centered, and plant subconscious memories of their time together. The importance of character and memory becomes key to this plot. Another salient example happens in the episode "The Time of Angels," as the Doctor leaves a blinded Amy alone in a forest filled with Weeping Angels. A few seconds after he leaves, he reappears on the screen and reminds Amy to always remember him. The audience may not take special note here, as the return of the Doctor to the scene is so quick and seamless, but, as revealed in "The Big Bang," the appearance of the second version of the Doctor was actually one from the near future, travelling back through time to save the universe.

On the other side of the Atlantic, other complex television shows also deal with characters finding themselves through temporal visits to their own memory. For example, in *Lost*, Desmond has the ability to travel in time and affect events in different time streams. Although hinted at in seasons two and three, Desmond's ability is dramatically revealed in the fourth season episode "The Constant," in which Desmond is able to send his consciousness back and forth between himself in the present and himself in the past. Desmond's ability is causing him to suffer great pain, and his inability to control it in the episode will eventually lead, as is so graphically demonstrated in the bleeding death of George (a fellow time jumper) to his death. The only way to stop it is if he can create a connection between his life in the "present" (2004) and his life in the "past" (1996). This connection, this titular "constant," is his girlfriend Penny, from whom in 1996 he is recently estranged and whom in 2004 he is trying to contact. Desmond's time jumping allows him to take information from 2004—Penny's phone number—and use it in 1996 to make that connection. Using knowledge gained in the past, Desmond is able to help himself stop time jumping and solve the problem in the present.

Importantly, the audience is as temporally confused as is Desmond in "The Constant," as the narrative follows the temporal jumps of Desmond as the metaphoric character. As represented on screen, *Lost*'s play with temporality illustrates the complexities inherent in today's online interactions. Transgenic media reflect both the past and the present at the same time. Users of transgenic media can "change" their representation of the past, and reflect a different sense of the "present" day. Through the reproducibility of online interactions, we live in a constantly shifting temporal sense. I can find out immediately what's happening this instant on Twitter, or can describe the past with unnerving detail. Our digital selves, mediations of our offline selves, exist in this interstitial space, between the *then* and the *now*; between the *past* and the *present*; between the *self* and the *mediated self*. Television reflects this type of temporal jumping using the metaphor of the time traveler as part and parcel of this sense of self-exploration.

Narrative Jumps in Characters

A second type of character temporal dislocation represents *narrative* jumps in time by following multiple storylines within a vast network of characters. Although I talk about this type of networking and its relation to online social network sites in greater detail in Chapter Six, it's important to note here that many complex narratives focus on this form of character temporality as a main function of the show's break with temporality in general.

As a form of 21st-century media, a focus on the complex interaction of vast networks of characters is a way for television to feature a contemporary aesthetic of transgenic media. For example, online social networking allows multiple viewpoints of a particular issue to become visually represented on one screen.[75] A controversial topic posted to your Facebook wall may become a sounding board for multiple perspectives from "friends" you forgot you had. As Jenkins has described, contemporary social media allow for a "collective intelligence" to form around areas of knowledge: no one knows everything, but by collecting the knowledge of humanity in one location (the Internet), we can all benefit from the vast collection.[76] Using the *metaphoric* dimension of characters, and focusing on narrative temporality and networking, contemporary television illustrates this type of collective intelligence. *Lost* serves as a useful bridge between these two different types of character temporality because it not only features characters that jump in *time* (Desmond, as we've seen, but other characters time travel in *Lost*'s fifth season) but also a narrative that jumps *characters*.

Specifically, the multitude of characters in *Lost*, and a narrative that follows one main character in each episode, means that certain events are revisited within the narrative framework of the show. And upon each revisit, new information is revealed or new character motivations are uncovered. As clear examples of this, the characters of Nikki and Paulo, two survivors of the plane crash, appear in the third season of *Lost*. Originally, the characters were conceived of by the producers in order to detail what some of the less-crucial characters were doing on the island.[77] But Nikki and Paulo also exist to help elucidate characteristics of the other protagonists, as well as serve as entry points for the audience to become more involved in the goings-on on the island. Nikki and Paulo are, in other words, producer-created audience surrogates that interact with the main characters. In the episode "Exposé," Nikki and Paulo's backstory is revealed in flashbacks to their time earlier on the island. For example, their first flashback details the time period immediately after the plane crash. Nikki, dazed after the crash and wandering around the fuselage of the plane, runs into Boone, who asks her if she has a pen. Sharp viewers will remember that Jack Shepherd, the doctor, had stopped Boone performing mouth-to-mouth resuscitation incorrectly, and Boone had (in a fit of quackery) decided to find a pen in the first episode to perform an emergency tracheotomy (Jack did not allow it to happen). The reveal in "Exposé" is telling, as it allows the audience to see the minute connections between characters, and to revise the meaning of previous plot points.

As audience-surrogates, therefore, Nikki and Paulo also function as metaphoric characters. Since they are both secondary characters who observe the action from afar, they parallel the viewing stance of the audience. Additionally, their presence in the show as ancillary to the main action serves to underscore the importance of the protagonist's action: Nikki and Paulo are the receivers of action, not the actors. It is telling that when they are mistakenly buried alive in their finale episode, "Exposé," Nikki's eyes pop open at the end: even as she dies, she is watching the action of others.

Although short lived as characters, both Nikki and Paulo interact with other characters in their flashbacks, allowing the audience to revise or flesh out its understanding of previous points in the *Lost* narrative. For example, a flashback sees protagonist Kate arguing with two characters who had died at this point in the series, Shannon and Arzt. This flashback details a scene from the same time frame as the first season episode "Whatever the Case May Be." In that early episode, Kate and Sawyer found a case of guns on the island. By introducing the characters of Nikki and Paulo into the previously depicted narrative moment from the show, the producers of *Lost* facilitate a revised understanding of the temporal situation of the narrative. Nikki and Paulo inscribe new meaning into an old episode. By traversing the network of characters across various time frames within the narrative of the show, then, *Lost* jumps temporalities by way of character relationships.

Two years before the 2004 premiere of *Lost*, a show aired with a similar character trajectory. *Boomtown* also features multiple perspectives on a single event. The seven main characters—two police officers, two detectives, a politician, a paramedic, and a reporter—all encounter a crime from different view points. Individual one-time characters in each episode also feature their own perspective, each of which revises portions of the narrative that have already been revealed. As the narrative of each episode unfolds, each character's particular perspective is revealed, uncovering the larger mystery. For example, in the second episode of the series, "Possession," the audience is privy to a phone call from a telemarketer, who overhears a murder plot on the phone. The audience is led to believe that it's a family dispute, when a husband confronts his wife's lover. Seeing the representation of the fight even further solidifies this response, as we see a pudgy older man aiming a gun at a young, muscular stud (a stereotypical male adulterer). Yet, further in the episode, and from additional characters' vantage points, the audience gains a deeper understanding of the situation: it is revealed that the older man is a john who frequents a sex club, and had hired the young woman to pretend to be his wife. Additionally, the aforethought adulterer is actually the real husband of the young woman. *Boomtown* revisits the scene where the older man aims the gun at the "adulterer" a number of times, and from a number of vantage points (the older man's, the young woman's, and the young man's). Additionally, the investigation is revealed through the detective's viewpoints, the crime scene through the officers' and the death through the paramedic's, each depicted out of linear order. Through temporal play, the meaning of the scene changes due to the different perspectives of the characters.

The science-fiction program *Fringe* also demonstrates the temporal displacement of characters as depicted by the "swapping" of bodies through alternate realities. In the show two universes exist in parallel to each other, and inhabitants with special abilities (including protagonist Olivia Dunham) are able to travel between the two universes. In the third season especially, the Olivia from the militaristic—yet prophetically familiar—universe replaces the Olivia from our own universe. The doubling of the character represents an ongoing struggle in the series, as male lead Peter, with whom our-Olivia has a flirtatious relationship, begins to date alter-Olivia, not knowing that she had been replaced. The idea that a doppelgänger exists that could replace us has a transgenic media parallel, as our profiles can mirror, but never replace, our offline identities. As van Kokswijk has shown, a created identity online can "exist independently from human control and can (inter)act autonomously in an electronic system."[78] That is, while we can control what we put on our profiles, we have no control over what others post, nor what corporations like MySpace or Facebook do with the information we put online. By creating alternate identities online, we are quite literally creating doppelgängers of our selves, articulating a personality online that may differ from what is presented offline. But we take that personality as real, as given: as Astrid remarks to our-Olivia when she returns to our universe, "[Peter] thought that she was you. Whatever feelings that Peter, had they were not about her. They were about you. And they were real. They still are."[79] So too are our profiles real; but they are also separate from us. Because our social networking profiles are our digital doppelgängers, when we experience an online connection between them and another digital persona, there is a fundamental separation between entities. Yet we still feel connected to it; violated if it is breached (as in Dibbell's famous depiction of a rape in cyberspace).[80] Television teaches us how to separate these two entities; how to survive in a digital alternate reality.

Furthermore, the show *Dollhouse* uses the personality-switching ability of the main characters to illustrate two sides of transgenic media's representation of identity—Jenkins's collective intelligence versus what Axel Bruns calls the "representations of knowledge." We use transgenic media everyday to record our personal history, which then becomes visible—and importantly—rewritable. Users of digital media "appear to be developing more familiar relationships with digital technologies," and such familiarity breeds integration into our lives.[81] Collective intelligence shows the attempt to represent all of humanity's knowledge in one location. In contrast, Bruns' idea of "representations of knowledge" facilitates a vast restructuring of what constitutes knowledge itself. By illustrating multiple viewpoints on a particular personality—different perspectives networked together—the Internet allows for greater fluidity in how information, how sociality, is presented. And this form of social knowledge affects how people respond to others online.

First, *Dollhouse* uses each of the Dolls to represent collective intelligence incarnate. Each doll is programmed with the personality of different people. Protagonist Echo is housed in the body of Caroline. The Caroline/Echo hybrid, however, soon learns how to "save" personalities, storing them like so many volumes in a Kindle. As Caroline/Echo starts to take on multiple personalities, she also starts to teach the others how to do the same. Through the singular voice filtered through the collective intelligence of her personalities, Caroline/Echo is able to complete her missions and tasks, aided by a community inside of her.

Second, *Dollhouse* also uses Caroline/Echo as a way of demonstrating how the web records all memories as equally valid. In "Stop-Loss," Caroline/Echo is uploaded with multiple military personalities in order to infiltrate the Rossom Corporation. She can use whichever personality is most relevant to the situation. Similarly, the "representations of knowledge" available for users of Wikipedia appear to have a similar effect as Caroline/Echo—Wikipedia doesn't attempt to illustrate one particular Truth but rather different subjective experiences, all of which could be taken as truth.[82] Caroline/Echo doesn't have *one* objective personality, but rather different subjective personalities, all of which could be substituted for her "true" character.

Conclusion

In sum, in each of these shows, the jumps in temporality based on a metaphoric character's motivations function to offer the character a chance to facilitate a greater understanding of his/her own life through reflection, self-awareness, or comparison with others. Contemporary television shows use temporal displacement to teach the complex interactions users have within their transgenic media. Ultimately, temporal displacement becomes one more way that contemporary complex television narratives function as both a means for and representation of a fracturing of personal identity. Creeber describes this as one of the main functions of television, to represent a fluid, postmodern identity: "In a postmodern world, the notion of a fixed, authentic self is called into question, the breakdown of traditional subject positions (based on nationhood, sexuality, race and even gender) becoming increasingly unstable."[83] Television becomes a means to link character to audience, through the metaphoric dimension of characterization. Character displacement illustrates explorational transgenic media.

Both transgenic media and contemporary complex television content seem to mirror the same cultural force. As Castells puts it, the Internet "is an extension of life as it is, in all its dimensions, and with all its modalities."[84] We like having control over our data online, seeing our personal history, and focusing on our connections with others. Television gives us a representation of that desire by illustrating the connections individuals can have with their past. The personal exploration experienced by characters through temporal displacement reveals a personal history that is a revisable, yet stable, aspect of our lives. In the next section of this book, I will expand on this notion to show how temporal displacement at the narrative level in complex television reveals a similar dichotomy between stability and revision in our conception of cultural history.

Notes

1. See Maria Nikolajeva, *The Rhetoric of Character in Children's Literature* (Lanham, MD: The Scarecrow Press, 2002), viii, for a detailed exception to this, although she focuses specifically on children's literature.
2. Altman, *Narrative*, 12.
3. James Phelan, *Reading People, Reading Plots: Character, Progression, and the Interpretation of Narrative* (Chicago, IL: University of Chicago Press, 1989); also James Phelan, *Living to Tell about It: A Rhetoric and Ethics of Character Narration* (Ithaca, NY: Cornell University Press, 2005).
4. Altman, 10.
5. Kozloff; although the narrative is often explicitly part of the program as well, as in *Desperate Housewives* (2004) or *Arrested Development*; also see Gregory Currie, *Narrative and Narrators: A Philosophy of Stories* (Oxford, UK: Oxford University Press, 2010), 123–47; as well as more generally Mieke Bal, *Narratology: Introduction to the Theory of Narrative*, 3rd ed., trans. Christine Van Boheemen (Toronto, ON: University of Toronto Press, 2009); Seymour Chatman, *Story and Discourse: Narrative Structure in Fiction and Film* (Ithaca, NY: Cornell University Press, 1978); Seymour Chatman, *Coming to Terms* (Ithaca, NY: Cornell University Press, 1990); Gérard Genette, *Narrative Discourse: An Essay in Method* (1972), trans. Jane Lewin (Ithaca, NY: Cornell University Press, 1980); Gerard Genette, *Narrative Discourse Revisited* (1983), trans. Jane Lewin (Ithaca, NY: Cornell University Press, 1988); Bordwell; Nikolajeva.
6. Roberta Pearson, "Anatomising Gilbert Grissom: The Structure and Function of the Televisual Character," in *Reading* CSI*: Television under the Microscope*, ed. Michael Allen (London: I.B. Tauris, 2007); Roberta Pearson, "Lost in Translation: From Post-Network to Post-Television," in *Quality: Contemporary American Television and Beyond*, ed. Kim Akass and Janet McCabe (London: I.B. Tauris, 2007); Roberta Pearson, "Chain of Events: Regimes of Evaluation and *Lost*'s Construction of the Televisual Character," in *Reading Lost: Perspectives on a Hit Television Show*, ed., Roberta Pearson (London: I.B. Tauris, 2009).
7. Horace Newcomb, "Towards a Television Aesthetic," in *Television: The Critical View* 3rd ed., ed. Horace Newcomb (New York: Oxford University Press, 1982); Robert Allen, *Speaking of Soap Operas* (Chapel Hill: University of North Carolina Press, 1985); Robert Allen, "Audience-Oriented Criticism and Television," in *Channels of Discourse, Reassembled: Television and Contemporary Criticism*, 2nd ed., ed. Robert Allen (Chapel Hill: The

University of North Carolina Press, 1992); Ivan Askwith, "'Do You Even Know Where This Is Going?' *Lost*'s Viewers and Narrative Premediation," in *Reading Lost: Perspectives on a Hit Television Show*, ed. Roberta Pearson (London: I.B. Tauris).

8 Pearson, "Anatomising."

9 Pearson, "Chain of Events."

10 This is obviously a generalization and there are exceptions: a well crafted anti-hero, like Tony Soprano, for example, can elicit just as much emotional connection as can the most kind-hearted soul. Identification doesn't have to be about being like or liking a character, but rather about having that character emotionally move an audience.

11 Newcomb, 486.

12 Chatman, *Story and Discourse*, 119.

13 Gaby Allrath, "Life in Dopplegangland: Innovative Character Conception and Alternate Worlds in *Buffy the Vampire Slayer* and *Angel*," in *Narrative Strategies in Television Series*, ed. Gaby Allrath and Marion Gymnich (Houndmills, UK: Palgrave Macmillan, 2006), 134.

14 Ibid., 137.

15 Shirky.

16 Foucault, "Self-Writing."

17 Creeber, *Serial Television*, 6.

18 See Ryan, 148–57.

19 Juan-José Igartua, "Identification with Characters and Narrative Persuasion through Fictional Feature Films," *Communications* 35, no. 4 (2010): 347.

20 Jonathan Cohen, "Defining Identification: A Theoretical Look at the Identification of Audiences with Media Characters," *Mass Communication and Society* 4, no. 3 (2001): 245.

21 See Joli Jensen, "Fandom as Pathology: The Consequences of Characterization," in *The Adoring Audience*, ed. Lisa A. Lewis (New York: Routledge, 1992).

22 Cohen, 251; emphasis mine.

23 Blakey Vermeule, *Why Do We Care about Literary Characters?* (Baltimore, MD: The Johns Hopkins University Press, 2010), xii.

24 Nancy Roser, Miriam Martinez, Charles Fuhrken, and Kathleen McDonnold, "Characters as Guides to Meaning," *The Reading Teacher* 60, no. 6 (2007): 548.

25 Cohen, 251.

26 Goffman, 252.

27 Newcomb, 486.

28 Bal, *Narratology*, 113.

29 Roser et al., 548.

30 Ibid., 548.

31 Ira Newman, "Virtual People: Fictional Characters through the Frames of Reality," *The Journal of Aesthetics and Art Criticism* 67 (2009): 73.

32 Ryan, 149.

33 Phelan, *Reading People, Reading Plots*, 2.

34 Eco, "On the Ontology," 83.

35 Ibid., 86.
36 Phelan, *Reading People, Reading Plots,* 2.
37 Eco, "On the Ontology," 89, 93.
38 Roser et al., 548.
39 Phelan, *Reading People, Reading Plots,* 2.
40 Newman, 74.
41 Phelan, *Reading People, Reading Plots,* 10.
42 Bal, *Narratology,* 112–33.
43 Nikolajeva, 8.
44 Newman, 75.
45 Ibid.
46 Currie, *Narrative and Narrators,* 49.
47 Murray, 214–47, although she doesn't call them metaphoric.
48 Ibid., 238.
49 Foucault, "Technologies."
50 Goffman, 252–56.
51 Ibid., 64.
52 See Bukatman.
53 Manuel Castells, *The Internet Galaxy* (New York: Oxford University Press, 2003), 118.
54 Bukatman, 9.
55 Ibid., 22.
56 Sobchack, *Screening Space,* 272.
57 See, in particular, Heather Marcovitch, "Memories of Mars: *Life on Mars* and the Discursive Practices of Memory," in *American Remakes of British Television: Transformations and Mistranslations,* ed. Carlen Lavigne and Heather Marcovitch (Lanham, MD: Lexington Books, 2011).
58 Jameson, 27.
59 Eco, "On the Ontology," 93.
60 Modleski, 68.
61 Booth, *Digital Fandom,* 157; David Horton and Richard Wohl, "Mass Communication and Parasocial Interaction," *Psychiatry* 19 (1956): 229.
62 A similar nostalgic interaction with show titles occurs in the third season of *Fringe.* When an episode takes place in the 1980s, as a few do, the titles reflect the time period, with 1980s-style videographic imagery and text.
63 David Lewis, "The Paradoxes of Time Travel," *American Philosophical Quarterly* 13, no. 2 (1976): 146.
64 Amanda Lenhart, Kristen Purcell, Aaron Smith, and Kathryn Zickuhr, "Social Media and Mobile Internet Use Among Teens and Young Adults," *Pew Research Center,* 2010, http://pewresearch.org/pubs/1484/social-media-mobile-internet-use-teens-millennials-fewer-blog (accessed 10 Mar 2010).
65 Creeber, *Serial Television,* 14.
66 Ibid., 15.
67 Cornel Sandvoss, *Fans: The Mirror of Consumption* (Malden, MA: Polity Press, 2005), 96.
68 Holdsworth, "'Television Resurrections'," 139.
69 An early example of this is *Quantum Leap* (1989), in which Sam Beckatt "leapt" in time, but could only travel within his own lifespan.

70 Nahin.

71 Greg Rickman, *Science Fiction Film Reader* (New York: Limelight, 2004), part 11.

72 Mary Madden and Aaron Smith, "Reputation Management and Social Media," *Pew Research Center,* 2010, http://www.pewinternet.org/Reports/2010/Reputation-Management.aspx (accessed 01 June 2010): 2.

73 Booth, *Digital Fandom*; also Jacob van Kokswijk, *Digital Ego: Social and Legal Aspects of Virtual Identity* (2007), rev. ed. (Delft, The Netherlands: Eburon Academic Publishers, 2008).

74 See Allrath.

75 Axel Bruns, *Blogs, Wikipedia, Second Life, and Beyond: From Production to Produsage* (New York: Peter Lang, 2008).

76 Jenkins, *Convergence Culture*, 26–29.

77 Askwith 2009; Cindy White, "Executive Producers Damon Lindelof and Carlton Cuse Promise That They've Found the Plot Twists That Will Bring *Lost*'s Viewers Back," *Sci-Fi.com,* 22 Jan 2007, http://web.archive.org/web/20070318060759/http://www.scifi.com/sfw/interviews/sfw14769.html (accessed 02 Mar 2011).

78 van Kokswijk, 13.

79 Alison Schapker, and Monica Owusu-Breen, writers, "Marionette," *Fringe* (Los Angeles: Bad Robot, 2010).

80 Julian Dibbell, "A Rape in Cyberspace: How an Evil Clown, a Haitian Trickster Spirit, Two Wizards, and a Cast of Dozens turned a Database into a Society," *The Village Voice,* 23 Dec 1993, http://www.juliandibbell.com/texts/bungle_vv.html (accessed Retrieved 14 Aug 2011).

81 Kirsty Best, "Living in the Control Society: Surveillance, Users and Digital Screen Technologies," *International Journal of Cultural Studies* 13, no. 1 (2010): 6.

82 See Bruns, 118–24.

83 Glen Creeber, "*Heroes,*" in *The Television Genre Book,* 2nd ed, ed. Glen Cleeber, Toby Miller, and John Tulloch (Houndmills, UK: Palgrave Macmillan, 2008), 57.

84 Castells, *Internet,* 118.

Section III: Temporalities

In this section I focus on temporal displacement as it relates to issues of time, narrative and history. In each of these chapters, I use the metaphor of "time travel" as a heuristic to examine how changes in our transgenic media use affect our everyday lives. In Chapter Four, I explore the idea of transgenic media as a permanent record of a personal history, and demonstrate its relation to contemporary narrative theory. An emphasis on history reveals the absolute *permanence* of information in the digital age, which is nearly indestructiblet. In Chapter Five, I contrast this view with one that shows transgenic media as deliberately ahistorical; that is we are in a perpetual present, constantly *updating* our digital lives. An emphasis on immediacy notes the fleeting and ephemeral nature of digital information, which can be manipulated with little or no traces left of the original.

I synthesize these two conceptions in Chapter Six, which argues for both historical and ahistorical views of transgenic media as a form of "networkization" through the lens of television studies. I contrast the networked nature of the science-fiction drama *Lost* with a similar "networkization" of time in the half-hour sitcom *How I Met Your Mother*. I argue that both shows represent a form of networked storytelling. Ultimately, in all three chapters of this section, time travel functions as a narrative device that saliently illustrates contemporary issues in transgenic media use. As both a way of fixing and fracturing history, time travel offers a heuristic that examines changes in our perception of time.

Four

Time Travel and the Codification of Narrative/History

Everyone has a time machine. Everyone is a time machine. It's just that most people's machines are broken. The strangest and hardest kind of time travel is the unaided kind. People get stuck, people get looped. People get trapped. But we are all time machines. We are all perfectly engineered time machines, technologically equipped to allow the inside user, the traveler riding inside each of us, to experience time travel, and loss, and understanding. We are universal time machines manufactured to the most exacting specifications possible. Every single one of us.

—Yu, How to Live Safely in a Science Fictional Universe, 165–66

The "time machine" about which Yu speaks in the above quotation is personal, individual, and uniquely human. In a very real way, we all time travel everyday. Yet, we don't necessarily think of the forward momentum of our progression through life as "time travel," simply because we have little to no control over this movement. As I write this, I *am* travelling in time; I am simply moving at a one-to-one ratio of experienced time to external time. At the rate of one second of my experiences, one second occurs in "real-world" clock time. As Lewis writes, this does not a classic time traveler make, as we assume that time travel:

> involves a discrepancy between time and time. Any traveler departs and then arrives at his[/her] destination; the time elapsed from departure to arrival (positive, or perhaps zero) is the duration of the journey. But if he[/she] is a time traveler, the separation in time between departure and arrival does not equal the duration of his[/her] journey.[1]

As I have explored in the first section of this book, even if the physical body does not time travel, the mind can imaginatively transport the individual into different times: our very human ability to think outside time gives us a form of mental time travel.[2] Obviously, this type of "time travel" deserves to be put in quotation marks, for it's not *physical* travel in any sense of the word. In his *Science Fictional* novel, however, Yu and his father have found a way to harness this imaginative potential of the human mind to travel in time by mechanizing it with a temporal machine that actually transports the physical body to times that the mind envisions.

In this chapter, though, I explore a more traditional approach to time travel and illustrate what happens when characters physically travel in time. What I will show in this chapter is that the representation of time travel on television serves as a facilitating heuristic for understanding contemporary, complex narrative. While our transgenic media produce new means of "mental" time travel, I want to extrapolate from this to examine the televisual representation of the theoretical and philosophical dimensions of physical temporal displacement.

Transgenic media *reinforce* a historical presence through the preservation of personal details. Ultimately, time travel narratives on television illustrate a structural connection to the historical record generated by transgenic media. Given the near-unlimited recording potential of digital data online, highly digitized individuals can be tracked and their histories understood far more completely than at any other time in the past.[3] Information can be stored permanently and recalled instantaneously. We live in a world surrounded by the remnants of our history. Is it any wonder that with so much history at our fingertips, our fictional televisual media have also started to concentrate on the (re)emergence of historical detail as a key factor in their narrative development?

Because we live in a world that thrives on causes and their requisite effects, we take it as rote that causality, like memory, is solely uni-directional: that is, that causes lead to effects. When I drop a mug of tea on the floor, it will break—it doesn't work backwards so that if I were to catch shards of porcelain that rose off of the ground I could then quickly reassemble them into a device that could hold hot liquid. In linear narratives, as Bordwell has shown, every aspect of the narrative becomes unified and "subordinated to the movement of cause and effect" within the narrative structure.[4] Yet, what we've seen in the past decade's worth of television is that the relationship between causes and effects is much more complicated than linearity first suggests. In temporally complex narratives, the existence of a complicated linearity leads to the development of narrative paradoxes. I believe an increase in the number of depictions of narrative paradoxes represents an aesthetic and textual reaction to transgenic media. It's not that paradoxes didn't exist in time travel stories before the 21st century, but rather that many shows today *thrive* on the paradox as a key narrative element.

In a narrative paradox, effects are often shown before, and often don't neatly sync up with, causes. For example, in the episode "White Tulip" of *Fringe*, scientist Alistair Peck has developed the ability to travel back and forth in time. With each jump, he sucks energy from whatever living things are near him—including, in the first minute of the episode, all passengers on a train car. The team of investigators discovers the dead bodies and investigates Peck, who then jumps back again to prevent being captured. This time, though, he alters his behavior to avoid the Fringe team. But savvy viewers might have wondered, if the second jump "negated" the first jump (as indicated by a shift in the Fringe investigation), then what happened to the *first*

Peck? The cause of his jumping would have been eliminated by the effect of his jump.

In literary theory, the paradox is a logical contradiction, a statement both *true* and *false* at the same time; for example, "This statement is false." Paradoxes in time travel mean a reversal of sorts, when one element contradicts another. For instance, the "grandfather paradox," perhaps the most common paradox depicted in the media, details the temporal contradiction that occurs when a time traveler goes back in time to before his/her own birth and kills his/her grandfather. How, then, has the traveler existed at all? In narratives, such paradoxes are extremely vexing. That is, because narrative is a *human* construct, it must be somewhat determined by human characteristics. The resolution of paradoxes becomes of the utmost concern, because *not* resolving paradox is troubling to conventional narrative theory, to cause and effect through our linear worldview. But, as Lewis points out, paradoxes are not *impossible*, merely *improbable* in our limited ideology.[5] Further, time travel physicist Paul Nahin shows that "paradoxes are often offensive only to human, culturally biased intuitions about how things 'ought to' work, since the classical physical laws are not offended by a reversal in the direction of time."[6] To reconcile a paradox means thinking in a dialetheisist manner, seeing that *both* the thesis and the antithesis of any proposition must be true. This kind of "thinking outside the box" in logical terms is difficult.[7] We dislike watching our narrative coherence disrupted.

Yet the paradox thrives today, as I will demonstrate by focusing on temporal complexity in the 2005 "New *Who*" revival of the BBC show *Doctor Who*. I want to make the argument, then, that a rise in narrative paradoxes in *Doctor Who* mirrors a similar rise in a paradox of transgenic media: namely, the *preservation* of material in online environments like Facebook and Twitter is matched with a simultaneous *instantaneousness* about our mediation. History, through temporal displacement is both recorded and erased through transgenic media, and this relationship becomes symbolized through *Doctor Who*'s temporally complex narrative.

Doctor Who and (Unstable) Narrative Temporalities

Doctor Who started in 1963 and ran almost uninterrupted until 1989. After a failed reboot effort in 1996 (an international co-production between the UK's BBC and the US's Fox network[8]), *Doctor Who* (as a television program) lay dormant for nearly a decade.[9] In 2005, British writer/producer Russell T Davies rejuvenated the series to critical and popular acclaim.[10] *Doctor Who* has become not only a successful television franchise, but also a lucrative transmedia event: today toys, comics, books, DVDs, games, hardware, costumes, and cookie jars join the television program in homes.

Doctor Who has managed to maintain a niche fan base while still appealing to a larger audience. For example, the final episode of the 2009 "specials" season, "The End of Time part II," garnered 12.27 million viewers in the UK alone.[11] The new series of *Doctor Who* seems to have been a success, surviving (and even thriving) despite two regenerations (from Eccleston's erratic Ninth Doctor, to Tennant's exuberant Tenth, and now Matt Smith's child-like Eleventh) and behind-the-scenes changes (Russell T Davies was replaced by Steven Moffat as producer in 2009). Yet, in many important ways, the new series is remarkably different from the old: gone are the cheesy monsters, in favor of flashy story arcs and computer graphics. Most significantly, however, there has been a shift on what the series focuses. Specifically, the notion of time travel has become arguably the central emphasis of the "New *Who*" series,[12] in contrast to the 1963–1989 Classic *Who* in which it was rarely dissected with anything other than cursory dialogue.[13] Although the Classic series does depict time travel, it is rarely explored in the depth the New *Who* series attempts—a few mentions of the *fact* of time travel usually served as plot constructs to propel the Doctor from story to story. In contrast, New *Who* uses time travel as a singular narrative device which presupposes a number of assumptions about the nature of time, history and memory, all of which *Doctor Who* skillfully—but playfully—negotiates and synthesizes through the travels of the Doctor and his companions.

I want to argue that of all the changes between the Classic and the New *Who*, the most dramatic is a concentration on multi-linear storytelling and dramatic shifts in narrative; this change is, I suspect, based largely on an audience which has grown to be comfortable with digital technology. In other words, while *Doctor Who* may not have created a generation of digital natives, it has altered its themes and narrative structure in order to embrace this population.

For example, contrast the difference in treatment of a temporal paradox between the 2010 episode "The Big Bang" with an equally complex temporal situation in 1975's "Genesis of the Daleks." In "The Big Bang," the final episode of the fifth of the new *Doctor Who* series, the Doctor, whilst completely encased in a giant vault called the Pandorica, also appears in front of surprised companion Rory, who recently watched the Time Lord get locked in the vault.[14] This newly appeared Doctor and the understandably shocked Rory have the following head-hurting conversation:

> Doctor: You need to get me out of the Pandorica.
> Rory: But you're not in the Pandorica.
> Doctor: Yes I am. Well, now I'm not, but I was back then... which is *now* from your point of view, which is *back then* from my point of view. Time travel—you can't keep it straight in your head.[15]

Rory's bewilderment is palpable, as the Doctor's appearance to Rory comes as a result of the Doctor's deliberate flaunting of cause/effect time mechanics. Namely, the Doctor travelled from the future, in which he *has* escaped from the Pandorica, to tell Rory how to help him escape from the Pandorica. Indeed, the nature of the Pandorica is fraught with temporal discontinuities. It broadcasts its message to everywhere and every*when* in time. This broadcast is picked up by enemies of the Doctor who travel back in time to Earth at this moment to capture him (in fact, we learn that this broadcast was sent out to trap the Doctor to this point). But up until this point in history, these events *had not yet happened*, even though it was history to everyone who witnessed it. Cause and effect have not just been reversed, but have actually been subverted.

While it would be an understatement to say that Rory is in complete shock to see the Doctor appear in front of him, the audience, attuned to the dynamism of time travel narratives throughout the temporally complex new *Who* series, may be more amenable to the intricacies of the time travelling titular hero. Importantly, the conversation between Rory and the Doctor is not just a *moment* in the *Doctor Who* text, but also represents a central theme throughout the series. In contrast, in the classic serial "Genesis of the Daleks," the Doctor philosophizes about his ability as a time traveler to erase the Daleks from existence by destroying them before they exist. The actual philosophy of time travel, however, is often ignored, slighted, or elided:

> Do I have the right? Simply touch one wire against the other, and that's it. The Daleks cease to exist. Hundreds of millions of people, thousands of generations can live without fear, in peace... and never even know the word "Dalek."[16]

No mention is made of the paradox this genocide would cause—and the Doctor's qualms are justified, for the ultimate destruction of the Daleks would not only "wipe out a whole intelligent life-form," but would also cause a Bilking Paradox, wherein the destruction of the Daleks would cause a break in the timeline so large that there would be no reason for the Doctor to go back in time to destroy the Daleks.[17] Instead of dealing with this paradox, the serial postpones the Dalek's destruction and wraps up the cause/effect narrative without exploring any of the larger time travel implications or causality loops involved in the process. In contrast to the minor temporal play of the "Genesis of the Daleks," the complex temporality of "The Big Bang" exists as an operational aesthetic that subverts the audience's conventional expectations for cause/effect timelines. This subversion becomes a major focus in the New *Who*. And *Doctor Who* is not alone in this: temporal displacement in narrative structure has become a common trope across a range of shows.

Temporality in Transgenic Media

The rise in popularity of the temporally complex narrative has occurred at the same time as the advent of transgenic media. I certainly don't want to indicate that one caused the other, but rather I believe that the emergence of both at the same time reveals a cultural focus on issues of complex temporality. Complex and temporally displaced narratives play with our notion of time; similarly, transgenic media also play with our contemporary notion of time in varied and multi-faceted ways.

There exists a sense of permanence within our transgenic media use that seems to belie the ephemeral sense of "time passing" that we experience in our off-line lives. To take just one macabre example, the existence of Facebook "death" profiles has helped illustrate how transgenic media emphasize the inerasable nature of our social software.[18] When a user of Facebook dies, there is, of course, no way for that person to erase their profile from Facebook's servers. Their disembodied virtual self (disembodied twice over, now) exists in the ether. And unless Facebook is notified that a user has died, they will continue to use that profile in a variety of ways: suggesting it to others who may want to "friend" it, connecting it on Google or other search engine sites, or encouraging others to "reconnect" if they haven't spoken in awhile. It is a shock, to say the least, when one is reminded by Facebook that it is your deceased friend's birthday and you need to write on their wall.

While some people may make sure a spouse or significant friend has their "log on" details and instructions to delete profiles upon death, many do not. And given the flurry of activity that occurs when someone dies, it may take many months before family members get around to Facebook. The end result is that our transgenic and digital persona still exists and continues to be "memorialized" in the digital terrain. And given that we as users have no control over who posts on our walls, or who contributes to our public persona, there are many weak-link "Facebook friends" who may not even *know* that the profile owner has died. Imagine waking up the day after a good friend has passed away, checking your Facebook news feed, and seeing that a weakly-tied but mutual friend has posted "Long time no see!" on your departed friends' page. The persona continues to live long after the person has died.

This is the paradox of transgenic media: to exist without existence; to provide a virtual space of insubstantial substance. The paradox of transgenic media is that it both *is* and *is not*. Despite their digital (i.e., intangible) existence, transgenic media are *things*—content that holds meaning—memorialized online forever. Our online self is persistent, as long-forgotten texts, images and videos can now be made permanent and, importantly, public on YouTube. The Presidential election of 2048 will be all the more exciting, as any candidate running will have lived the majority of his/her life with all their digital lives—their Facebook profiles, their YouTube accounts, their Google Searches—saved, indexed, and stored for future journalists and opponents to study.

Time takes on a different connotation with transgenic media as our sense of temporal expectations are changed. "New" media have always had this effect: we used to wait a month for a letter, now we expect replies to our text messages within minutes. Before cell phones, if a phone rang without being answered we assumed no one was home; today, if it clicks over to voicemail after two rings we know we've been "ignored." The permanence of our existence in the digital ether is made manifest by our numerous profiles online. Google's web searches save our online information, as anyone about to embark on a blind date, a job search, or a college application knows firsthand: what will they find if they "Google" you? But my digital presence, my online "self," isn't just constructed by me on Facebook, MySpace, Bebo or Orkut—rather, it's constructed from the bits and bytes of data that *others* have put online as well: my College's homepage, my high school sports activities, my former employer. Even if one has a common name ("Paul Booth" is particularly difficult to enact as a search term, given the popularity of my moniker doppelgänger, the world-famous tattoo artist), Google's search engines run deep and for thousands of queries.

Transgenic media reference and maintain a permanence of our lives, recording what has come before. Take, for example, the nearly ubiquitous Gmail account. According to social media researcher Jason Zalinger, Gmail reflects an ongoing obsession in our culture to record, preserve, and reflect upon our past.[19] We are "now all digital storytellers, historians and autobiographers of our own lives. And if this is true, for millions, Gmail is one of the primary tools used every day to record the bits and bytes of our narrative life."[20] Gmail is a permanent archive of our digital communication, and it stores our conversations, our calendars, our tasks, and our contacts forever. Yet, beyond permanence, Gmail also highlights and reflects our interaction with our narratives with others in our lives:

> Stories guide our decision making. If I [G]mail (I'm using [G]mail as a verb on purpose) you a story about a professor who is intolerable, you might not take his class. Of course, I might be completely wrong about the professor. Whether you take the course or not, my story becomes part of the story of your semester, and if you tell someone else, and that person [G]mails me about the professor then our Gmail narrative completes its communicative circle.[21]

Yet, importantly, Gmail as a transgenic medium is not just about one person: it necessarily takes two people to communicate asynchronously to create a Gmail narrative.[22] And, as if to reflect that, today's complex narratives use temporal displacement to underscore a similar duality at the heart of fictional narratives as well.

Temporality in Narrative

Narrative is the primary way human beings structure their understanding of their place in time. Narrative also creates a structured linearity, linked symbolically to our "real-life" experiences. As Ricoeur reminds us, all narrative takes place within a prescribed time, an "uncriticized temporal framework."[23] As his narrative philosophy describes, the main function of a narrative is to *construct* and *order* events in a time frame, and as readers we enter into a mutual agreement with the author to condense or expand the timeline. Time travel highlights this role by creating two separate but connected plots. In the first plot's timeline, a time traveler ventures outside his/her timeline. In the second plot's timeline, the series of events that the time traveler *leaves* structures its own linear experience, from which the time traveler is now removed. In *Doctor Who* terms the Doctor exists "outside" of time, but that doesn't mean that he himself is not affected by time—he ages, but he ages in a way counter to the rest of the (linear) universe. As an audience we follow his story, but the discourse could just as well illustrate Rose's, Sarah Jane's, Donna's, or Amy's.[24] It is in our (forced) "choice" to follow the Doctor instead of the linear universe that our experience of *Doctor Who*'s narrative becomes destabilized. And by following *his* life, the external events of the universe's narrative timeline are altered, de-ordered and presented in a multi-linear representation.[25]

Symbolically, the two time frames created via time travel are mirrored in all narratives, regardless of medium, by the dichotomy between the story and the discourse.[26] In classic narrative theory, the story represents the underlying events and characters "within the world of a show, whether they are shown on screen or not."[27] The discourse (often also called the *plot* of the show) describes the way that story is told, through the events actually on-screen, the choices the producers, actors and directors make, and the order in which the audience sees the events. Bordwell defines the story as "the action [of the narrative] as a chronological cause-and-effect chain of events occurring within a given duration and a spatial field."[28] In other words, although the story "is never materially present on the screen," it exists as the association viewers construct from the events as presented on screen.[29] What is viewed onscreen, the discourse, "is the actual arrangement and presentation" of the story.[30] In other words, the *story* describes what happens in a narrative, while the *discourse* represents how that story is told to the viewer. Like in a time travel narrative, what we see on the screen represents one time frame, while our mental reconstruction of that discourse into a chronological order represents another. The story/discourse divide is fundamentally about perception, and places the audience squarely at the center of narrative interpretation. The difference between story and discourse can best be summed up by examining a complex television show like *24* (2001–2010), a real-time action/adventure show which follows a multitude of characters. As viewers, we are shown only snippets of each character's day; what we see on-screen is the discourse. However, to reconstruct what happened to the characters in between the moments we see them is to re-create (or imagine) the story.

It should come as no surprise, then, that time travel in fiction offers a useful way of renegotiating what narrative scholar Richard Walsh calls the "complex relations between narration (whether authorial or not) and fictional representation."[31] Audiences take meaning from narratives and apply it to their own lives to learn about and discover aspects of their selves. For Walsh, this distinction, between what a narrative says and what an audience takes from it allows audiences to create their own meanings in a narrative. Time travel offers a mode of storytelling that highlights the divide between the narrative *story* and a narrative *discourse* by integrating the role of the audience in the connection between events.[32] The metaphoric relationship between the story and the discourse is thus one that is paradoxically both constructed and constructive.

One of *Doctor Who*'s most important contributions for a scholarly investigation lies in this exemplar treatment of changing narrative tropes.[33] Specifically, the audience follows the Doctor's timeline, and the underlying story of the (linear) universe is constructed from the audience's fragmented experiences with the Doctor. In this subjective frame of reference, the audience uncovers the story from the exposition of their time with the Doctor. But the Doctor acts and reacts *across all time*. Therefore, to understand time in this manner means that the audience must recognize that any decisions the Doctor makes *must have already been made*. Everything that has, will have, or continues to have happened, has already happened in the universe. The audience views what is "now" to the Doctor, but which is always necessarily "past" in the specific story of the universe. While the discourse of *Doctor Who* must therefore contain the events and elements that make up the story of *Doctor Who*, it is only through a combination of the linear and nonlinear telling of the discourse that the audience sees the story. As I described in Chapter Two, the 2010 Christmas special "A Christmas Carol" represents this linear/nonlinear discursive style, as the majority of the discourse takes place *now*, but the Doctor also travels to the past (which we see *now* in the discourse) in order to elicit changes in the present/future.

To time travel within a narrative presupposes the separation of discourse and story, but also hinges on the fact that these two descriptors are inextricably linked. This logical duality lies at the heart of *Doctor Who*, and indeed of any time travel narrative. We can see this dual nature of narrative in "The Pandorica Opens" and "The Big Bang." The Doctor has just learned that the Pandorica—a mythical prison which was built to trap a wizard ("the Pandorica? Ha! That's just a fairy tale," exclaims the Doctor in "Flesh and Stone")—turns out to be his own prison, and he is the "wizard" in his own Time Lord fairy tale. Surprised at this revelation, he screams, "No! Please listen to me! The TARDIS is exploding right now and I'm the only one who can stop it!"

Indeed, the timing of the opening of the Pandorica in the year AD 102 matches the precise point in the discourse time at which the TARDIS disappears in the future (AD 2010—the palindromic match to the year of the Pandorica's opening). Through cross-cutting, producers of *Doctor Who* indicated to the audience that these two events were linked because they were visually represented simultaneously. But when exactly is "*now*" for the Doctor in the story? These two points, separated by thousands of years, are linked in the discourse, as the *now* in the Doctor's sentence seems to be referring not just to a linear progression of time, but also to a construction of time as *simultaneously co-existent* linearities. This temporality is mirrored in the relationship and the dichotomy between the Doctor's narrative story and discourse. We see the events as simultaneous *and* causal, even if in linear time they are not. Yes, the TARDIS is exploring now, but it also explored nearly 2000 years ago. Along this same theoretical line, Lewis argues that "time must be not a line but a plane."[34] The relationship between the Pandorica and the TARDIS is the resolution of a paradox, as each gives the other meaning. This bifurcation of meaning, this contradictory yet inseparable divide in narrative theory, is reflected back at the audience, for the audience sees both events happening simultaneously in the discourse, although of course the opening of the Pandorica occurred two thousand years prior to the explosion in the story.

Similarly, the Doctor was sent back to that point in time *precisely because* the Pandorica was opening. Cause and effect—the central tenets of narrative itself—are thrown into confusion, as the opening of the Pandorica signaled the end of the Universe precisely because the Pandorica was opening in order to prevent the end of the universe. The Pandorica's opening is timed perfectly from this one event: the explosion of the TARDIS and the locking up of the Doctor. The only way this could happen is if we imagine that time travel in *Doctor Who* assumes that time is multi-linear and that all points emanate from an ever-changing "now," moving in multiple directions equally (in the way we measure BC in a numerically similar, but directionally opposite manner than we do AD). What *Doctor Who* implies, therefore, is not that time is uni-linear but that multiple time frames exist *at the same time.*

As Hills suggests, the use of time in *Doctor Who* may be "more fundamental to the series' narrative construction" than anything else.[35] For a show like *Doctor Who,* which projects contemporary issues into science-fiction settings, the way it deals with narrative complexity reflects an increased temporal complexity in our own lives.[36] Therefore, just as we can examine the narrative of *Doctor Who* to elucidate issues of narrative sustainability, so too should we examine the implications of complex narratives on our quotidian selves. The increasing complexity of the time travel narrative of *Doctor Who* emerges during a period in time when our own subjective lives are under constant scrutiny. In short, we "time travel" through our personal narrative every time we glimpse our Facebook news, our RSS feeds, our wiki trends, etc. Like Sardick in "A Christmas Carol," we literally watch our immediate past/present flow before our eyes, affecting our present/future. In his authoritative book about time travel, Nahin illustrates this flow by quoting H. B Franklin: "'when one says time travel what one really means is an extraordinary dislocation of someone's consciousness in time.'" [37] It is this dislocation which is so significant to today's complexity in terms of the *Doctor Who* narrative.

Our use of transgenic technology has only increased this sense of timelessness in our lives: according to Andrew Gorman, "there is a pervasive uneasiness about our present and an uncertainty about our future, along with a concurrent nostalgia about our past" all functioning at the same time in our contemporary media usage.[38] Take, for example, the use of "tabs" on today's Internet browser windows. Instead of opening a new browser window for every webpage one wants to view, browsers such as Internet Explorer, Apple Safari, Mozilla Firefox, and Google Chrome allow the use of tabs: multiple windows open on the same browser. One could simultaneously open a Facebook page, a Google search, an email program, a Netflix Queue, and so on. Each of these windows forms a multi-linear experience of the web; each opens onto a particular time and a particular space. Indeed, the use of these hypermediated windows interacts with our own experiences online, such that each window appears to have a different time frame. We may not be time travelers like the Doctor while we sit at our computers, but we peek into different time frames, different temporal situations. Our single browser becomes a vehicle for the exploration of the temporal displacement of our selves. My *linear* experience of these media does not reflect the *multi-linear* existence of the browser's windows; but these windows reveal a *multi-linear* slice of my life. Like the Doctor, we can return to the past and "refresh" the present. I certainly don't mean to insinuate that *Doctor Who* is the *cause* of our hypermediated life, nor do I indicate the opposite; rather, I merely contend that the paradoxes at the heart of New *Who* become symbolic of our narrativized experiences as "a particular way of meaning, a particular kind of contribution to cultural discourse at large."[39] *Doctor Who,* in this way, is both an exemplar of, and a comment on, quotidian changes in our cultural life.

The Paradox and the Resolution

To synthesize these changes, *Doctor Who* relies on the paradox. By presenting time as a complicated tangle that obfuscates causes and effects, the paradox remakes narrative linearity into a knot. As Alec Charles states, the New *Who* series resolves paradoxes using causal loops, which represent a closed temporal system.[40] For example, a causal loop occurs when characters from the future are inspired to go to the past by an event that they themselves caused during their trip.[41] The "Bad Wolf" trope from the first New *Who* series is a quintessential example: Bad Wolf Rose travels back in time to leave the phrase "Bad Wolf" at various temporalities during her travels to spur her past self to become Bad Wolf Rose.[42] Instead of languishing within an unstable portrayal of time, New *Who* relishes the *stability* engendered by a well-produced paradox, as it represents a fully multi-linear view of temporality.[43]

To recognize this stable set of historical markers within our cultural history requires the ability to see multiple viewpoints at once. Shows like *Lost* and *Eureka,* for example, depict multi-linear narratives, where a changed event in the past doesn't necessarily conflict with present time frames. Instead, by changing events in the past one can create *new* presents, new linearities, which become alternate realities that don't necessarily erase the original reality. History is not changed; it is made. A multi-linear view of time travel symbolizes a view of time as flexible. The paradox is a malleable experience shaped by the subjective experiences of the human being through their own life narrative.

For time travel narratives, paradoxes are *de rigueur* but audiences have come to expect a resolution, no matter how "gobbledygook" it might be. A perfect example of this comes in the *Doctor Who* episode "Blink," in which the Doctor explains the complex temporal mechanics of time travel: "People assume that time is a strict progression of cause to effect, but actually from a non-linear, non-subjective viewpoint—it's more like a big ball of wibbly wobbly, timey wimey, stuff." Does this make sense? Only sort of…but it only *sort of* has to in order to make the narrative make sense.[44]

A key example of a resolution to a temporal paradox in *Doctor Who* is the character of River Song. Just as the Doctor doesn't traverse time in a linear fashion, neither does River Song. However, her time travelling is opposite the Doctor's: whenever they met, she is younger and he is older. Her appearances, although few, have had momentous effects on the Doctor. Specifically, River Song first encounters the Doctor at some point in time *after* the adventures presented in the fourth and fifth New *Who* series. Yet, the Doctor first meets her at the end of her life, as she dies at the end of the episode "Forest of the Dead."

River Song reconciles paradoxes into an inherent shape by understanding a dialetheistic answer. In the penultimate episode of the fifth series, Amy Pond and River Song have a conversation that explores this fundamental instability in human thought:

> Amy: This Pandorica thing. Last time we saw you, you warned us about it ...
> River: Spoilers.
> Amy: No, but you told the Doctor you'd see him again when the Pandorica opens.
> River: Maybe I did. But I haven't yet. But I will have.[45]

In fact, River Song ties together these issues of temporal paradoxes within *Doctor Who* and the narrative structure I mentioned earlier in the chapter. By correlating between points in time and points in a temporal structure (which are two different things in the *Doctor Who* universe), River Song bridges the divide between time frames, between discourse and story, and between memory and paradox. Her existence is both paradoxical (how can she meet the Doctor before knowing him?) as well as a resolution to paradoxes (she notified the Doctor about the Pandorica in the past after learning about it in the future, and her time travel reconciles the Doctor's temporal instability). In this way, River Song symbolizes the contemporary complex narrative: temporally displaced yet narratively consistent.

Such temporal instability in complex narrative is inherent to any sort of time travel story, but it is especially prevalent in one so characteristically dependent on paradoxes to maintain narrative consistency like *Doctor Who*. Given the nearly fifty-year history of the show, it is natural for the show to revisit and reevaluate central concepts of the narrative. But few shows have engaged so spectacularly with narrative revision as has *Doctor Who*. From changing the lead actor, to unexplained new interiors of the TARDIS, to revisions of canon information (did the Doctor steal the TARDIS? Or did he build it?), to completely rewriting the history of the Time Lords: it's surprising there's any stability at all.[46]

The exception that proves the rule is the first episode of the second season's "The Space Museum," one of the few stories of Classic *Who* that deals with time travel as something more than primarily as a means of getting from (time)point A to (time)point B. The First Doctor and his companions Ian, Barbara and Vicki materialize in an alien museum but on a different "time track" from the rest of the world. To explain this, Vicki uses "gobbledygook" to describe the makeup of time: "Time, like space, although a dimension in itself, also has dimensions of its own."[47] The four adventurers come face-to-face with themselves as exhibits in the museum, "waiting for time to catch up so they can join the story."[48] The time travelers have landed slightly *ahead* of their own future, and realize that they are actually "out of synch" with the normal timeline, as they leave no footprints, make no sounds, and hear nothing in this alien world. But instead of traveling back in time to correct their error, they merely wait for time to "catch up" with them. They have leapt into a future but are traveling backwards through it, although time itself is still "normally" flowing toward the future. Imagine travelling across North America by leaping into the air and waiting for the Earth's rotation to transport the rest of the landmass before landing. And the audience is watching them *progress* while time *regresses* until both timelines merge. If they've landed in the future but are traveling backwards through it, how can the audience view them normally? But then why do the *rest* of the characters also move forward? The instability of time becomes apparent, and the episode begs the question, in what direction does time flow?

As I noted at the start of this chapter, there is a significant difference in the way the New *Who* deals with time travel and paradoxes as opposed to the Classic series. Indeed, *Doctor Who* author and expert Robert Shearman explains in a documentary that "very few [classic] *Doctor Who* stories ever tackled time travel properly."[49] The unstable time travel plot, however, is not mentioned in the second, third, or fourth parts of the "The Space Museum" story. Classic *Who,* in deferring to narrative coherence, symbolizes a modernist approach to cultural consistency, where events follow a consistent path and paradoxes are impossible to maintain. In contrast, the paradoxes that lie at the center of the New *Who* series symbolize the inherent multi-linearity of a postmodern age. "Blink" represents a meaningful exploration of this multi-linearity. The Weeping Angels, the main villains in "Blink," have the power to send people back in time and feed off the potential energy of their victim's not-yet-lived life. Four characters end up being sent back in time: the Doctor, Martha, and Billy wind up in 1963 while Kathy Nightingale travels to 1920. The Doctor must somehow find a way to communicate from 1963 to the inhabitants of 2007 and get them to send his TARDIS back to him. But how to communicate across the span of forty–odd years? His solution is ingenious in its complexity: by encoding hidden messages onto DVD Easter eggs, the Doctor simulates a conversation with the recipient of these DVDs in 2007, his friend Sally, whose collection of DVDs spurs the Doctor to come up with this plan. And how does the Doctor know what list of DVDs to encode, or what to say in this half-conversation? Why, Sally gives him this list and a transcript of their conversation in her future (2008)—which, given the Doctor's time travelling, is actually *his* past.

Discontinuities between the story and the discourse abound in "Blink," which the smoothness of the dialogue and mediated communication elide. In fact, communication is key in "Blink," but much of this communication comes in the displaced form described in the previous chapter as the "Book from Nowhere." The DVD in the episode represents this displacement viscerally. To Sally, watching the Doctor on the TV, it appears as though she is speaking with a recording. But because the Doctor recorded one side of the conversation from the notes Sally gave him in the future, the recording actually moves as a normal conversation would, using the "horological mechanism" of past recording and present dialogue.[50]

The script written by Sally and given to the Doctor at the end of the discourse (but the beginning of the story, from the Doctor's perspective), becomes a form of Bilking Paradox, in which the construction of an element crucial to the time travel story seems to appear out of nowhere. Sally writes up her notes about the incidents to send to the Doctor in her future so that he can receive them before she encounters those events. But she transcribes the notes from her "conversation" with the Doctor, during which he reads a copy of her transcript. If Sally hadn't written the document the Doctor wouldn't have known what to say; but Sally only knows what to write because of what the Doctor says—so where does the document come from, "originally"?[51] Similarly, at a structural level, the play between discourse and story uses the paradox to cohere the narrative together. Coincidences that might seem phony in a traditional narrative—Sally happens to leave the DVD shop just as a movie-gazing clerk mutters to himself, "go the police!"—work in this time travel story to hold the narrative together. The "temporal disorientation" of "Blink" gives the show an "emotional realism."[52] And by playfully recognizing this paradox between the story and discourse—the Doctor's flippant "wibbly wobbly, timey wimey stuff" satisfies by obfuscation—the show relishes these two crucial components of a complex television narrative; and, therefore, shows how *discontinuity* can structure a cultural history just as easily as can historiographical continuity.

Doctor Who, however, is not the only show that uses temporal displacement to illustrate narrative discrepancies. In the first season of the television show *Heroes,* for example, protagonist Hiro is displaced to a different time frame in numerous ways, of which not the least significant is his doubling as Future Hiro. In the episode "Hiros," a Hiro from the future, differentiated from present-Hiro by the possession of a sword, the ability to speak English, and the presence of a fierce soul patch, tasks Peter with finding Claire. Later in the series, Future Hiro appears again, warning other characters of imminent disasters or crucial plot changes. In the fourth season episode "Once Upon a Time in Texas," however, temporal confusion occurs when Present-Hiro travels to a past that he had experienced once before (in the episode "Seven Minutes to Midnight") and sees himself (as the audience did) interacting with Ando. Past-Hiro recognizes Present-Hiro, but thinks that it's Future-Hiro instead (he asks, "where's your sword?"). A moment unseen in "Seven Minutes to Midnight"—Hiro in the restroom—is fleshed out in "Once Upon a Time in Texas," as the audience learns that Present-Hiro (posing as Future-Hiro) convinces Past-Hiro to travel back six months to save his girlfriend, Charlie (later depicted in the episode "Six Months Ago"). *Heroes* creates temporal confusion for audience and character alike.

The Hiro-Displacement confuses the narrative by creating two time frames, both of which the audience must keep track of—the narrative of the show, with humans learning to cope with the emergence of superpowers, and the narrative of Hiro's personal time frame. Events from one time frame occasionally appear in the main narrative, but largely the two co-exist separately. Additionally, paradoxes occur between the two, as Charlie's death causes Hiro to act in the first season, but the later prevention of this death would seem to indicate a reversal of this motivation.

Ultimately, the notion of the paradox becomes crucial for understanding the metaphor of discontinuity in temporally displaced television narratives. Hiro's separate time frames mashup to create one complex narrative. In the New *Who* series, there is nothing more dangerous than changing the past because one is actually shifting a narrative. Rose encounters such an incident in "Father's Day," when she saves her father's life in 1987 by averting his fatal car crash. The paradox affects all of time and space, as changing this event causes temporal disruptions across a range of technologies: Rose's mobile phone picks up messages from Alexander Graham Bell and a 1987 car stereo music plays music from the future. In retaliation for the disruption in time, the Time Vortex (an entity from which time sprang) sends "Reapers," giant dragon-like creatures, to devour all regions of temporal instability. In *Heroes*, the change is less violent and more innate: preventing the future means the death of Future-Hiro but that doesn't stop him from attempting it.

In tandem, there is also nothing more dangerous to the stability of a coherent narrative than a logical paradox in a discourse. If a narrative doesn't hang together, or doesn't make sense, it can ruin the audience's immersion in the story. The notion of the paradox as a plot device is eschewed as the instability this brings in the temporal depiction of the discourse clouds narrative coherence. The paradox, in both narrative and time travel, becomes the end result of the interaction with time as the participants explore it. And precisely because paradoxes are narratively untenable, they become adversarial in time travel narratives. The quintessential example of this appears within the two-part finale to the fourth New *Who* series, "The Sound of Drums/The Last of the Time Lords." The Master constructs out of the Doctor's TARDIS a "Paradox Machine" that allowed the Toclafane to invade and destroy the "normal" timeline of Earth. The Toclafane are the evolutionary "next step" for humanity; they are what we become. The Master convinces them to enslave the humans of their own past. Of course, since this didn't actually happen in their/our past, their invasion becomes a paradox. The Master uses the stolen Doctor's TARDIS to create a ("gobbledygook") Paradox Machine that keeps this causal loop open, and allows the Toclafane and the Master to rule present-day Earth from an alternate reality. From a narrative theory standpoint, the notion of a "Paradox Machine" and an alternate universe are necessary not only for the plot to conclude, but also in order for narrative coherence to be saved. If the machinery of paradox removal doesn't exist, the world of *Doctor Who* would have been irreconcilable with the narrative events as the audience knows them, and complex narrative coherence would be lost.

Yet, there remains one paradox, fully articulated in the finale of the New *Who* specials ("The End of Time") that continues to plague the coherence of the overarching narrative of the show: the erasure of the Time Lords and the Daleks in the Time War. The paradox of the disappearance of the Time Lords is vast and problematic from a narrative standpoint. Namely, it raises the fair and complicated question of, what happened to Susan?[53]

Susan, the Doctor's granddaughter, had been a traveling companion for many years when the series started. If, as the Doctor maintains about the Great Time War, all the Time Lords (and Daleks) had been removed from existence when the war ended, then how does he have a Granddaughter? Is the past Classic *Who* moot? Or does it exist in an alternate time frame? Ultimately, the questions are unanswerable, because the *Doctor Who* narrative asks us not to ask them, usually through verbal wordplay or elision that belies the complex narrative confusion underlying the concept. For example, in "The Beast Below," Amy asks the Doctor why he looks human. He responds:

> Amy: So there's other Time Lords, yeah?
> Doctor: No. There were, but there aren't... just me now. Long story. It was a bad day, lots of bad stuff happened.[54]

The Time Lords have disappeared from all of time and space, yet the effect of their existence is still felt in the universe. As an audience, we must accept this fact, while also accepting that, say, the existence of Rose Tyler's father in the universe for a few moments longer than he "should" have existed is cause for the destruction of all. Or that if old Billy contacted young Billy in "Blink" that two thirds of the universe would be destroyed.

On the one hand, we understand that this is a children's television show, and that such complexities can (and should) be written off so that we can get on with the adventure. But on the other hand, in a show that, at least recently, has depicted so often time travel explanations and paradox reversals, it appears out of place that this fact remains unexplained. Perhaps part of the confusion over these unexplained elements lies in the fact that so much *is* explained. "Blink" has a fully articulated time travel concept; "Turn Left" and

"The Last of the Time Lords" illustrate alternate world theory in a narratively satisfying way. Yet, the "End of Time" leaves off with more questions than gobbledygook answers.

Time Travel and Narrative Complexity

As we've seen, time travel offers a mode of storytelling that highlights the divide between the narrative *story* and the narrative *discourse*, and the influx of television narratives focusing on time travel underscores the quotidian familiarity with fragmentation in our media and our lives. By fracturing the boundaries we have become used to in our culture, we not only expand our ability to think beyond the immediate, but we also teach ourselves about living in a fractured life. This boundary-breaking mirrors the structural divide implicit within what Jason Mittell has termed "narrative complexity" in contemporary television programming, a term that inherently brings with it temporal complexity.[55]

One aspect of narrative complexity particularly relevant to temporal displacement is narrative moments that carry over not only from episode to episode, but from television series to series. Specifically, the narrative complexity of a franchise like *Star Trek* represents a visceral exclamation of temporal displacement. In the 1980s, the *Star Trek* franchise got a "reboot" with *Star Trek: The Next Generation* (1987), and followed in the next decades by *Star Trek: Deep Space Nine* (1993), *Star Trek: Voyager* (1995), and *Enterprise* (2001). While none of these shows concentrated on serial storytelling in the same way as *Lost* or *Flashforward* did, they did continue character development and some narrative information from one series to the next. For example, Captain Jean-Luc Picard in *Star Trek: The Next Generation* was kidnapped and temporarily turned into a Borg, cybernetic life forms bent on galactic domination, and "Locutus of Borg" was also mentioned in *DS9* ("The Emissary") and *Voyager* ("Scorpion, Part II"). The Borg War, which happened towards the end of *ST: TNG*, had repercussions felt throughout *DS9* and *Voyager*. Further, individual character narratives continue throughout the first four series, as protagonists from one show often appear in others.

Worf is a pristine example, as he was a main character on *ST: TNG* during its seven year run, and then became a main character on *ST: DS9* during its final four seasons. Individual narrative moments connect the series as well. For example, a 1968 episode of *Star Trek* (the original series) featured *The Defiant,* a ship which had been captured by Tholians and trapped in an unstable area of space-time between universes ("The Tholian Web"). In a 2005 episode of *Enterprise* (a prequel to the original series), *The Defiant* appears in the alternate universe, leading to increased hostility with the Tholians.

Enterprise, in fact, marks one of the first instances of long-form narrative in the *Star Trek* universe. Although it premiered last of all the franchise's series, it takes place before any of them, including the original series. As a prequel, then, *Enterprise* occupies a unique temporal moment for *Star Trek*: it originates much of the information the audience later sees in the shows, but uses that origination as a form of conclusion as well. Thus, when we first see the Borg in the *Enterprise* episode "Regeneration," it is with the full knowledge of their malicious behavior and the war to come in the previous series. Further, the Borg in *Enterprise* exist in a temporal loop themselves, as they had come to Earth in the past in the film *Star Trek: First Contact,* and had been frozen for the century that took place between the film and the episode. The presence of the Borg in the past before humanity knew of their existence is of course an example of a causal loop. As described on *Memory Alpha,* the online wiki devoted to *Star Trek* (I quote at length to illustrate the complexity of temporal mechanics here):

> At the end of ["Regeneration"], it is revealed that the assimilated Earth freighter dispatched a subspace message to the Delta Quadrant...implying that this may be how the Borg Collective originally learned of Humanity's existence....If the message was received, it may explain why, in the 24th century, the Borg specifically wanted Captain Picard to speak for them and why they sent at least one vessel to the vicinity of Federation space, destroying several Romulan and Federation outposts in 2364. (TNG: "The Neutral Zone") ...A Borg sphere travelled to the 21st century in an effort to assimilate Humanity in the past. (*Star Trek: First Contact*)

> The sphere was destroyed, and several drones from the vessel were frozen in an Arctic glacier in 2063, which were uncovered by the Arctic scientists 90 years later in "Regeneration," starting the causality loop all over again.[56]

This form of connection—the networking of information from one storyline to another—mirrors two important facets of contemporary television programming. First, as I've described in this chapter, the way the audience must reconstruct the temporal flux of the *Star Trek* franchise helps teach the skill of reconstruction necessary to piece together similar fragments in our everyday lives. And second, the complex interweb of temporal relationships in the franchise highlights the networkability of the show, a topic which I will discuss in more detail in Chapter Six.

Conclusion

In late 2011 Facebook released its "Timeline" feature. This new layout actively rearranged users' profiles so that events were listed chronologically, on a seemingly endless timeline of a user's life. Everyone's timeline starts on the day of their birth and, depending on their overall usage of Facebook, continues to add information throughout the user's own personal timeline. For example, my timeline extends from my birthdate through my graduation from high school (according to Facebook, that's the next important event in my life—sorry 5th-grade production of *Julius Caesar*, I guess no one's posted about you yet), and continues through such momentous events as my marriage, my current employment, and my pet adoptions. Yet, Facebook's timeline itself reveals an intricate user-generated paradox that mirrors the sort of narrative paradox engendered by *Doctor Who* and other temporally complex narratives. The story of my life is not necessarily mirrored by my own discourse about my life. I can reflect back in time, posit future ideas, and even post photographs of myself from years past. Unless someone alerts Facebook to these discrepancies, my discourse does not match my

story: my life is a "wibbly wobbly, timey wimey" Mobius strip, stretching forward and backward from an ever-present *now*.

I believe that this common networkability across texts, across media, across narratives and across characters finds solid reference with the way contemporary media engender both mutability and stability in their users. Postmodern theorists Fredric Jameson and Jean Baudrillard have both argued that our contemporary culture has flattened any distinctions between past, present and future.[57] To these theorists, we live in an extended present made permanent only through our mutual recognition of how the past reflects in the present and how the future springs from it. We are the most wired, most visible and most connected beings in our history. Our lives are stored electronically in segments across the digital landscape. Our personal life narratives can be read and reread throughout time. Through a culture of immediacy, we receive our instant media gratification purely digitally but can observe and preserve it indefinitely. Facebook and other social networks allow us to externalize our selves online.[58] I can, at any time, view my profile and experience a snapshot of my life at a particular moment in time. I am, in effect, seeing myself as I want to be perceived, in a time frame of my choosing, permanently. Similarly, in fifth series' "The Hungry Earth," the Doctor, Amy and Rory arrive in 2020 and see the future Rory and Amy standing across a field from them. The future Rory and Amy seem to be visiting the location where they know their past selves have been (where the current Rory and Amy now stand). The future Rory and Amy tour a moment from their past, revisiting their former selves, as if I were to search through my Facebook posts from 2006 and revisit my life then.

Narratively complex shows like *Doctor Who* are characterized by a type of complex *audience* that similarly finds narrative enjoyment out of the duality of off-line and on-line life. This focalization plays off the temporally discontinuous zeitgeist of the era. The end result of this audience reflection is that narrative reflects life just as life reflects narrative: What *Doctor Who* seems to be interpreting here is what Sean Redmond states about our current culture, that "the modern world produces a particularly acute identity crisis and existential

schizophrenia"[59] Previous *Doctor Who* series' played off the cultural factors when they were made.[60] The difference today is that the new series seems to rely on the paradox as a structure in order to relate to a multi-linear view of our lives, which revolves around technology and media use. Time travel gives us the narratological tools necessary to investigate the way discourse and story play out in a complex narrative. In a complex world, audiences need both stimulation and familiarity. Mittell references this in the dialogue between episodic and seriality, as the most complex narratives are those that maintain an episodic familiarity but generate serial speculation.[61] Time has to function in two paradoxical but related ways in *Doctor Who*. Time has to be fixed and constant. In "The Big Bang" the Doctor must be able to travel to the writing on the cliff face. Importantly, he must also be aware that the cliff has writing on it before he goes. But time also has to be more mutable and less constant than anything else. Events have to be able to happen in the past that have not yet happened before, even if participants in those events have no memory of them happening (that is, that they happen *for the first time* in the past even if the past events have already transpired to the point that they have become legend). The writing on the cliff must be untranslatable until the precise moment it can be translated.

This temporal complexity may appeal to audiences not just because of *Doctor Who*'s history, but also because of its relation to our everyday life. We watch *Doctor Who* to help linearize our non-linear lives; to give meaning to the meaningless; to see solutions to what we perceive as paradoxes. Ultimately, the time travel narrative on television becomes a way to look both at and through our contemporary lifestyle of hypermediation.[62] Only by reconciling the paradox and the multi-linear narrative can we see the quotidian life experience of the contemporary digital citizen. Returning to "The Big Bang," we can see how *Doctor Who* becomes a way to reify this concept. When the Doctor appears and speaks to Rory, even though he is also still trapped in the Pandorica, he has doubled his own time line, folding himself back upon himself. Two of him now exist, one encased in a mythological box created out of someone else's memory,

the other one about to show Rory how to release himself. Just as we can recover pieces of our selves from our memory, or from our digital archives, so too can time travel help us here. As the Doctor earlier (later?) says in "The Big Bang":

> People fall out of the world sometimes, but they always leave traces. Little things we can't quite account for. Faces in photographs, luggage, half eaten meals. Rings. Nothing is ever forgotten, not completely. And if something can be remembered, it can come back.[63]

Throughout our use of transgenic media, we can control our personal history. But there is also a certain comfort we take with the permanence of the multi-faceted presentation of information online. Indeed, just as the multi-linearity of narrative emerges from the interaction of the story with various discourses (or the discourse with various stories), so too do we create multi-linear discourses of our own lives structured via online networks. What *Doctor Who* and other time travel narratives present to us is a paradox function, which works precisely *because* of the impossibility of stability in transgenic cultures. Instead of reacting against the variability contemporary society, these time travel narratives interact with, augment, and reinforce a multi-faceted contemporary quotidian life.

Yet, to reinforce the instability of culture is just one function that *Doctor Who*'s time travel narrative creates. In the next chapter, I want to take a more specific look at the relationship between *Doctor Who* and a contemporary fluidity of historical "truth" through an examination of tardis.wikia and the *Doctor Who* narrative. Much like the fluidity of transgenic media articulates a need to rethink much of what we know about narrative, so too does it articulate a need to rethink what we know of history.

Notes

1. Lewis, "Paradoxes," 145.
2. Thomas Suddendorf and Michael Corballis, "Mental Time Travel and the Evolution of the Human Mind," *Genetic, Social, and General Psychology Monographs* 123 (1997): 136. Their idea that time travel can be a trope to aid our human development mirrors the relationship between time travel and narrative development nicely.
3. See Bruce Sterling, *Shaping Things* (Cambridge, MA: The MIT Press, 2005).
4. Bordwell, 157.
5. Lewis, "Paradoxes," 145.
6. Nahin, 135.
7. See Laurence Goldstein, "A Unified Solution to Some Paradoxes," *Proceedings of the Aristotelian Society* 100 (2000).
8. See Hellekson.
9. *Doctor Who* never truly died, though—fans kept the show alive, and scores of books, audio dramas, and graphic novels were released during the 1990s and 2000s.
10. See Matt Hills, *Triumph of a Time Lord: Regenerating Doctor Who in the Twenty-First Century* (London, UK: Taurus, 2010).
11. See the Broadcasters' Audience Research Board, http://www.barb.co.uk/; and according to Robert Siedman, "BBC America Starts 2010 With Best Ever Night and Day," *TV By the Numbers,* 06 Jan 2009, http://tvbythenumbers.zap2it.com/2010/01/06/bbc-america-starts-2010-with-best-ever-night-and-day/37825 (accessed 11 Apr 2011), the US broadcast was the most successful showing on the little-viewed BBC America cable channel, garnering the channel's high of 1.335 million viewers. This is not counting digital downloads (both legal and illegal).
12. See Hills, *Triumph.*
13. Important exceptions exist, which I detail in this chapter and the next.
14. For ease of reference, when it is not obvious, I will refer to Classic *Who* as stories within seasons, and will refer to New *Who* as episodes within series (see fn 59).
15. Steven Moffat, writer, "The Big Bang," *Doctor Who* (London: BBC, 2010).
16. Terry Nation, writer, "Genesis of the Daleks," *Doctor Who* (London: BBC, 1975).
17. See Nahin, 134.
18. See Dan Fletcher, "What Happens to Your Facebook After You Die?" *Time,* 28 Oct 2009, http://www.time.com/time/business/article/0,8599,1932803,00.html (accessed 5 May 2011).
19. Jason Zalinger, "Gmail as Storyworld: How Technology Shapes Your Life Narrative" (PhD diss., Rensselaer Polytechnic Institute, 2011).
20. Ibid., x.
21. Ibid., 6–7.
22. Although with G-chat it is possible to communicate synchronously as well.
23. Paul Ricoeur, "Narrative Time," *Critical Inquiry* 7, no. 1 (1980): 170.
24. In "Blink" the audiences does follow Sally Sparrow linearly, but the complexity increases with her non-linear interaction with the Doctor.

25 See Ken Dancyger, *The Technique of Film and Video Editing,* 3rd ed. (New York: Focal Press, 2002), 392.

26 Seymour Chatman, "What Novels Can Do That Films Can't (And Vice Versa)," *Critical Inquiry* 7, no. 1 (1980): 122.

27 Jason Mittell, *Television and American Culture* (New York: Oxford University Press, 2010), 217.

28 Bordwell, 49.

29 Ibid.

30 Ibid., 50.

31 Richard Walsh, *The Rhetoric of Fictionality: Narrative Theory and the Idea of Fiction* (Columbus, OH: Ohio State University, 2007); 6.

32 See Chatman, *Story and Discourse*; Genette, *Narrative Discourse.*

33 See Rob Shearman and Toby Hadoke, *Running Through Corridors: Rob and Toby's Marathon Watch of Doctor Who* (Des Moines, IA: Mad Norwegian Press, 2010).

34 Lewis, "Paradoxes," 145.

35 Hills, *Triumph*, 105.

36 Ibid., 91; also, Marc Edward DiPaolo, "Political Satire and British-American Relations in Five Decades of *Doctor Who*," *Journal of Popular Culture* 43, no. 5 (2010); Alec Charles, "Ideology of Anachronism: Television, History, and the Nature of Time," in *Time and Relative Dissertations in Space: Critical Perspectives on Doctor Who*, ed. David Butler (Manchester, UK: Manchester University Press, 2007).

37 Nahin, 7.

38 Andrew Gorman, "*Back to the Future:* Oedipus as Time Traveller," in *Liquid Metal: A Science Fiction Reader*, ed. Sean Redmond (Middlesex, UK: Wallflower Press, 2004), 116.

39 Walsh, 6.

40 Charles, "Ideology of Anachronism," 119.

41 See Nahin, 205–14.

42 "Bad Wolf Rose" is Rose's name after she absorbed the Time Vortex.

43 Nahin, 193–201; and Hills, *Triumph,* 88; see also Charles, "Ideology of Anachronism," quoted in Hills, *Triumph*, 87–89.

44 See Hills, *Triumph,* 98; quoting Stanislaw Lem, "The Time Travel Story and Related Matters of Science-Fiction Structuring," *Microworlds* (London: Mandarin, 1991): 146.

45 Steven Moffat, writer, "The Pandorica Opens," *Doctor Who* (London: BBC, 2010).

46 See Lance Parkin, "Truths Universally Acknowledged: How the 'Rules' of *Doctor Who* Affect the Writing," in *Third Person: Authoring and Exploring Vast Narratives*, ed. Pat Harrigan and Noah Wardrip-Fruin (Cambridge, MA: The MIT Press, 2009).

47 Glynn Jones, writer, "The Space Museum," *Doctor Who* (London: BBC, 1965).

48 Robert Shearman, "Defending the Museum," documentary on DVD, *Doctor Who: The Space Museum* (London: BBC, 2010).

49 Ibid.

50 Hills, *Triumph*, 98.

[51] See also Richard Hanley, *Is Data Human? The Metaphysics of Star Trek* (New York: Basic Books, 1997), 201.

[52] Hills, *Triumph*, 99–100.

[53] I am indebted to my student Courtney Neal for first voicing this question.

[54] Steven Moffat, writer, "The Beast Below," *Doctor Who* (London: BBC, 2010).

[55] Mittell, "Narrative Complexity"; David Lavery, "*Lost* and Long-Term Television Narrative," in *Third Person: Authoring and Exploring Vast Narratives*, ed. Pat Harrigan and Noah Wardrip-Fruin (Cambridge, MA: The MIT Press, 2009) calls them "flexi-narratives."

[56] "Regeneration (episode)" *Memory Alpha*, 2010, http://memory-alpha.org/wiki/Regeneration_(episode) (accessed 22 Dec 2010).

[57] Jameson; Baudrillard, "Ecstasy."

[58] See Booth, *Digital Fandom,* 151–80.

[59] Sean Redmond, *Liquid Metal: A Science Fiction Reader*, ed. Sean Redmond (Middlesex, UK: Wallflower Press, 2004), 115.

[60] See DiPaolo.

[61] Mittell, "Narrative Complexity."

[62] Bolter and Grusin.

[63] Moffat, "The Big Bang."

Five

Time Travel and the Production of Narrative/History

We have evolved to have current, temporally proximal beliefs about the world...which is to say local-scale accurate beliefs...We perceive the present but we remember the past. The converse is not possible. We obviously cannot remember the present. Or can we? Déjà vu. What does that feel like? It is the oddest experience, one everyone has had, one that is commonly described as a feeling of certainty that one has experienced just this exact experience before.

—Yu, How to Live Safely in a Science Fictional Universe, 179

For the type of time travel in Yu's novel, the past is a stable entity that we can visit but can never change, where there is a sense of finality, and where what has happened will always have happened. For example, when Yu the character revisits his past, all he can do is watch. The past is a fixture in time, and interaction is simply impossible—if it didn't happen in the past, then it obviously can't happen in the past even if the past is currently present.[1] Yu's mention of déjà vu, however, highlights a paradoxical circularity to his argument. *Now*, the present moment that is currently being inscribed

in history, is *going* to be past—instantaneously. Indeed, even in the very utterance of the word "*now*," we have already passed that present and that "now" is past. Are you the same person who just read, "For the type of time travel in Yu's novel, the past is a stable entity"? You have also changed since that "now," moments ago, much like I have changed since I just wrote it. There can be no present, for at the very instant we think "present," it is past. We shake off this feeling of asynchronicity not because it is meaningless, but precisely because it comes loaded with meanings we don't want to have to face.

Rewriting the past, for example, is only impossible when we look at the past from the present point of view. But from the past's point of view, writing what is to come is as easy as you or I deciding to change our routine for the day—to play hooky on a pop quiz and fail instead of pass. The linearity of our own experience of time locks us into one particular temporal frame of reference. Because we live our lives from cause to effect, and because we do not have any physical time machines, we are stuck with an unchangeable past. In the previous chapter, I described time travel as a metaphor from which we can draw much insight about the nature of narrative. The definition of a time traveler is one whose personal time is different from an external time frame, and I showed how the depiction of traveling to a different period in time can problematize the notions we typically take for granted about the flow of narrative. I also argued that this metaphor of time travel, as witnessed in contemporary complex television programming, can teach us to shift our understanding of our everyday life, symbolized through the *preservation* of historical norms by transgenic media.

In this chapter, I want to discuss the opposite view, not to counter my previous argument, but to highlight the simultaneous duality of transgenic media. Transgenic media illustrate both history and a loss of history in fluid and variable ways. Even though our transgenic media reinforce the historical context of our lives by continually recording and preserving our personal history, they also allow us to concentrate on the immediate present over all other stimuli. Our

conception of history itself has changed with this new digital paradigm (as Baudrillard describes, history "has become its own dustbin"[2]). For example, Twitter is continually updating, and thus becomes a nearly effortless way to keep an eye on what is happening *now* in the world, as the spread of the news about bin Laden's death demonstrates.[3] Search your Tweetdeck for how rapidly updates arrive and how much they change the focus of your feed. Or go on Facebook to see a continually updating and changing list of what's *present* in your community's lives. We refresh our Facebook profiles every five minutes to see what's "new"—and if nothing is, we impatiently insist on updating our own status. If you are not online you are not *now*. In this view of our relationship between our transgenic media and our history, we can see an elision of the historical in favor of the immediate. Even as I write this, my email buzzes with an incoming message, my Facebook alerts me that I have a new friend request, and my *New York Times* alerts sends me a message about today's political argument. History is unstable with transgenic media precisely because it's forgotten as soon as something newer comes along.

I want to undertake a more nuanced and detailed examination of the relationship between time travel on television and transgenic media in this chapter, but specifically focus on how *Doctor Who* presents temporal displacement as a way of illustrating *historicity*, or the notion that history itself is a construction of continually updating and always-present cultural and narrative meanings. By representing shifts in notions of history—what I term the Meta Code, the Diegetic Code, the External Code, and the Future Code—*Doctor Who* illustrates the flexibility and the instability of history both as sign and as a mode of narrative. *Doctor Who*'s use of history exemplifies the way knowledge and history function together online. Specifically, I examine wikis as a way of eliding and flattening the historical record. Wikis represent a particularly salient view of how memory, culture, and history collide. By illustrating an important tension present within editable digital archives, wikis represent how the archive of knowledge is both a historical misrepresentation as well as a necessary step in equating "knowledge" with historical truth.

Although transgenic media may seem to be perfect examples of democratic, technological public forums, texts like wikis instead exemplify the beliefs, ideas, and moments of the past that are inherently political, partial and problematic.[4]

In *Doctor Who,* temporal uneasiness is endemic to the Doctor's travels. As he emphatically states, "I've got a complex life. Things sometimes don't happen to me in the right order. Especially weddings. I'm rubbish at weddings...especially my own."[5] Time, for the Doctor, is an inherently non-linear experience, a truth he must learn to accept as a Time Lord who both creates and encounters temporal displacement in a variety of places. As viewers and, indeed, from the Doctor's own perspective, we see that the Doctor lives linearly. We know the First, Second, Third Doctors as they follow each other chronologically. But from an omniscient perspective, the Doctor does not—literally, cannot—live his life in a linear fashion. Yet linearity is the most inherent structure of the *human* experience, mediated or not. Trying to reconcile the linearity of our life (and the show *Doctor Who*) becomes a struggle when faced with inherently non-linear experiences (the Doctor's life; time travel storytelling).

Indeed, the way the Doctor describes *time* in space could just as easily be the way cultural historians refer to *history* on the quintessential historical text online, the wiki. Wikis, websites that facilitate the communal writing, editing and deletion of information, represent a view of knowledge as fluid and collective. Further, as a corollary to this collective behavior, wikis also provoke a sense of knowledge not as stable and indelible, but rather as *always already being rewritten*. *Doctor Who* offers an emblematic look at the way wikis entwine history and narrative. For, not only does *Doctor Who* deal with the mutability of time as an irresistible, inescapable force, but so too does the *Doctor Who* narrative reconceptualize the notion of history. The inevitable alterability of this history/narrative connection comes to a head in the fan-created wiki for *Doctor Who,* <*tardis.wikia.com*>.[6] Wikias are advertising-supported wikis that anyone with an Internet connection and email address can create about any subject he/she wishes. Wikias are often written about

popular culture texts, including the immense *Lostpedia* for the television show *Lost,* the detailed *Wookiepedia* for the *Star Wars* series and the immense *Megaman.wikia,* a site dedicated to the Nintendo video game series *Megaman.*[7] As encapsulated through the existence of *Doctor Who*'s *tardis.wikia,* the production of knowledge on wikias becomes an inherently social activity. It also, importantly, adds a deliberate sense of *mutability* to knowledge production.

I believe that an understanding of knowledge production on wikis—as representative of knowledge production in general—can be represented through an analysis of the historical/narratological tension inherent in *Doctor Who*. Further, as represented by *tardis.wikia,* an analysis of the fictional history in the *Doctor Who* narrative parallels the understanding of historical knowledge in contemporary digital environments.[8] I contend that *Doctor Who,* as a popular culture text, as a long-form television narrative, and as a prime example of temporal displacement, provides a useful inductive heuristic for understanding the fluid nature of knowledge in our everyday lives.

Although obviously not a factual science program, and the actual mechanics of time travel would not be as exciting to most viewers as the latest Dalek design or Sontaran invasion plan, *Doctor Who*'s focus on time travel is central not only to its narrative, but also to its sense of history. I begin this analysis with a description of wikis as representations of knowledge and a discussion of *Doctor Who* as a representation of narrative. Following this, I describe the historical codes on *Doctor Who* and establish how each one features in the companion wikia. Finally, I examine wikis as a form of narrative history, as indicative of the immediacy that transgenic media engender.

Historicity, Wiki Theory and *Tardis.wikia*

History can be interpreted as a construct of the particular time period in which is it made, and as more than a static element of the particular *past* from which it emerges. This is not to say that there's no such thing as the *past,* but rather to assert that our particular understanding of what the past means is constructed by our current social/cultural

situation.[9] We call this construction "history," and as Hayden White has asserted, we view these "putatively real events" through a narrativized lens.[10] That is, the way we understand history is not through our understanding of what *actually* happened, but through stories that describe those past events as if they were a narrative. When there is no narrative, there is no history.[11]

Problematically, however, if we examine our understanding of history from a postmodern standpoint, we find that the nature of history changes significantly. Famously, Fredric Jameson has noted that our postmodern era has found "a new depthlessness...both in our relationship to public History and in the new forms of our private temporality."[12] According to Jameson our past has been externalized from our immediate experience. That is, through mediated representations of history, we become obsessed with what Caldwell calls "the aura of history," or the notion that history is disembodied from our collective lives.[13] We've entered a realm of "historicity," where history itself is "excessive and stylistically overcoded," and understood as mediated television and cinema.[14] The media present aspects of history to us, but since we have no immediate context with which to understand the validity or authenticity of the mediatized events, we use that representation as a referent upon which to judge authenticity in general. Ultimately, as Robert Rosenstone suggests, "visual media have arguably become the chief carrier of historical messages in our culture."[15]

That's not to say we forget that we're watching a representation of history and think the events to be authentic to the time period depicted. As a case in point, Eckart Voigts-Virchow notes that the temporally complex British sitcom *Blackadder* (1982) becomes "history" for everyday viewers: "the next generations's historical knowledge is to some extent being shaped by a television sitcom [*Blackadder*], which has... contributed to a postmodernist, metahistoric social construction of history."[16] *Blackadder* plays on the "indisputably constructed nature of our representations of the past" in order to highlight the "constructivst paradigm of metahistorical thinking."[17] In other words, while no one confuses *Blackadder* with historical *truth*,

viewers have no other basis for judging what "truth" could possibly mean given the dearth of historical knowledge in our culture.

But this is nothing new: the idea that narrative shapes history has pervaded hundreds of epistemological paradigms, and the fact that we receive our history from the movies or from television certainly doesn't rock the boat pedagogically.[18] What I hope to uncover in this analysis, however, is how the representation of history within contemporary television narratives affects our understanding of *historicity*. The fact of a mutable history is nowhere more obvious than in transgenic media, where minute changes to an individual's profile can alter previous employments, erase unhappy relationships, or efface historical detail.

For instance, this historicity is perfectly encapsulated by wikis, which can be both source and storage of narrative information, although they portray narrative in a form different from what we are used to seeing in traditional media. Wikis present narrative not as beginning-to-end linear causation, but rather as separated events, each hypertextually linked with what I have previously called a "narractive" organization.[19] The ability of users to continually revise the wiki highlights the fact they are constantly impermanent. A wiki can never end: it is entirely unstable, a process not a product. It incites and invites change, and engenders "an interactivity that depends not just on the ability to change something onscreen, but also on the ability of users to communicate with each other."[20] As Bruns points out, wikis exemplify the diverse sets of knowledge bases that make up a community's multi-faceted viewpoint on a subject.[21] It is not that Wikipedia, for example, attempts to show "The Truth," but rather that it idealistically shows many different truths that complement or contradict each other to create a complete knowledge base. Because wikis can be continually updated, they are inherently present, but also remain tethered to ideas about the past: in this way, they are, quite literally, multiple representations of what Crary argues are the "different meanings [that emerge] depending on how [texts are] situated historically."[22] Wikis allow a space where truth can be

rewritten, where knowledge is changeable and where history is literally written by anyone with a web connection.

And this is why *Doctor Who* as a narrative text may be perfect fodder for the wiki. Just as information on wikis can be revised at a moment's notice, so too does historical information on and of *Doctor Who* go through a continual process of rewriting and revision through many different iterations.[23] Tulloch and Alvarado's classic work *Doctor Who: The Unfolding Text* illustrates how the show, at least in its first twenty years, may have attempted to maintain the *idea* of consistency rather than actual canonicity, as symbolized by the regeneration process.[24] The same can be said for *narrative* on the television show *Doctor Who*. Narrative itself is a continual process that runs both backwards and forwards in time; audiences must decipher elements of narrative in order to make sense of elements that appeared both earlier and later in the series. *Doctor Who* takes this narrative concept one step further: by actually playing with time as an inconsistent, yet constant series of events, the concept of narrative truth is interrogated in *Doctor Who*. For example, the Doctor can affect [has affected?] historic events and the program has, in fact, shown audiences such major historical moments as the rise of the Aztecs, the French Revolution and the burning of Rome. As moments in the *Doctor Who* universe, they stand beside and astride what viewers know to be "authentic" history. As a fictional television program, *Doctor Who* makes no claims of authenticity, and no one (hopefully) watches *Doctor Who* thinking that he or she is watching actual historical re-enactments. But as a representation of a particular type of *narrative* truth, *Doctor Who* forces a reconceptualization of the nature of history with every episode. The Doctor is not just a time traveler, but also a revisionist historian; so too is the wiki not just a website, but also a revisionist narrative.

Wikis preserve the specifics of an event, but can alter or change them if the historical truth is reviewed: if, for example, we learn the *actual* burning of Rome occurred differently than we first thought, our specific knowledge would change and "history" would literally have to be rewritten. Textbooks and primers the world over would require

months, if not years, to be revised. Wikis require only seconds. And yet although wikis are supposedly "democratized" media, in that anyone can write and publish, each writer is necessarily constrained not just by his/her cultural background, but also by the technologically determined format of the text. Although "anyone" can edit "anything" on a wiki, one must always follow certain rules, regulations, and syntax. "Anything" is a constrained term.

Wikis do not record and present history, then, but rather are in the process of constantly recording a *narrative* of history. This sense of narrative is difficult to see when looking at historical events on Wikipedia, as our immediate knowledge about history is colored by the way it is already narrativized online. In this way, wikis come to represent in practice the same ideological practices that have shaped and molded our conceptions of history. In a concrete sense, wikis provide a tangible illustration of the shifting relationship between time and power. This idea of a constantly updating temporal location for history and knowledge contrasts with what Debord discusses in his *The Society of the Spectacle*. For Debord, all knowledge is circumstanced around particular moments in time, discursive arenas of mediated power. The modern "spectacle"—the over-capitalization of the image as it dominates the masses—placates the population through "a paralyzed history…a paralyzed memory, of an abandonment of any history founded in historical time."[25] As Crary suggests, "Debord sees the core of the spectacle as the annihilation of historical knowledge—in particular the destruction of the recent past. In its place there is the reign of a perpetual present."[26] The spectacle, today's über-mediated environment, becomes a tool used to control the masses.

I want to argue, however, that wikis demonstrate a different type of "presentism" than what Debord originally argues. Although wikis-writers may mirror the ideological underpinnings of today's society, the underlying strength of the wiki is the way it is continually present in its relationship to history. History is not stolid on a wiki, but rather in a constant state of rewritability. There is no "annihilation" of historical knowledge; rather, a persistent "refresh" of historical details

as they pertain to particular units of knowledge. Wikis do not make consumers passive, but rather aid in the activity of the meaning-producing audiences. Debord sets the "innovation" that is "ever present in the process of the production of things" at odds with "consumption," which "is never anything than more of the same."[27] Through their own production, wikis demonstrate a *presentism* which is concerned not with erasure, but with accumulation.

We can see how this representation functions more precisely if we examine the development of an actual narrative as if it were history on a wiki. For example, watching the development of the narrative of the premiere episode of *Doctor Who*'s fifth series, "The Eleventh Hour," can offer insight into how narrative functions on wikis. Before the premiere, little description of the episode existed on *tardis.wikia*, and only two sentences constructed the plot:

> The Doctor has regenerated into a brand new man, but danger strikes before he can even recover. With the TARDIS wrecked, and the sonic screwdriver destroyed, the new Doctor has just 20 minutes to save the whole world—and only Amy Pond to help him.[28]

Yet, barely 21 hours later when the episode aired, the plot had expanded nearly nine-fold, and two days after that had blossomed into a full-fledged synopsis, complete with descriptions of scenes and characters that had been added since the first iteration. Over one year later, the plot summary has grown to nearly 2,000 words and lists the myriad connections that became apparent only as the series progressed.[29] Obviously, it is difficult to generalize from the description of one particular page on *tardis.wikia*. But what is remarkable here is the sheer transience of the narrative contained on the wiki. The narrative is constantly being performed in the present tense, despite depicting past referents.

As mentioned above, historiographers like White have noted that history itself is just a narrative we tell ourselves to link events. Indeed, as White points out, narrative is a "meta-code," the "basis of which [transmits]…messages about the nature of a shared reality."[30] In other

words, an historically grounded discussion of culture must necessarily be based in narrative terms. We create the concept of "history" to order and make sense of an ultimately senseless progression of events. And online hypertextual writing may be one of the most effective ways to display this historicity. As historian Stephen Robertson notes, "History (as a subject) is the ideal vehicle for hypertext—the myriad connections between subjects can be visualized and actualized more completely."[31] At the same time, *Doctor Who* and other fictional narratives are also written, more obviously so in this case. History, in both senses, is always fluctuating. By allowing users to edit, convert or rewrite entire sections of history, the wiki tangibly represents the mutability of historical truth, by "using links to create the more dynamic relationship between historical interpretations and the evidence on which they are based."[32] It is not that wikis are representations of historical absolutism, any more than *Doctor Who* is. For example, as Beer and Burrows show, wikis are particularly credited with a "new rhetoric of 'democratization.'"[33] These stories and images of the web circulate in scholarly and popular press: the way that "Web 2.0" opens up discussion of "collaborative, participatory or open culture."[34] Yet, as they go on to show, these discussions of the democratization of the web elide the fact that much of the web is "commercial and [a] lightly regulated market."[35] What is written on wikis is not always truth, and it's not always universal.

Historio-Narrative Truth

"Time is being rewritten," concludes the Doctor, towards the end of "Flesh and Stone," the fifth episode of the 2010 series of *Doctor Who*.[36] "Time," he follows up, "is being unwritten." The Doctor intones these words to River Song, the character who, as pointed out in the previous chapter, represents a nexus of temporal distortion. The Doctor, in this quotation, is referring to the supernatural disappearance of *time* as an entity in and of itself. Indeed, at the conclusion of the 2010 series, history literally is erased from ever happening—a fact made salient by the setting of the majority of the

final episode in a rapidly disappearing National History Museum. In a science-fiction show like *Doctor Who*, such occurrences are par for the course; and for the Doctor, as a time traveler, such utterances are commonplace. In the previous chapter, I described them as examples of "gobbledygook," placed in the narrative in order to "explain away" a paradox. But taken as a statement of knowledge, such gobbledygook actually has a profound effect on how we understand the concept of time in *Doctor Who.* Time as a concept is often described in *Doctor Who* not as an objective phenomenon—observable but unalterable—but rather as an immediate, subjective experience; one that can not only be changed but can actually influence events themselves.

Time, for the Doctor, is always unsettled. For example, he often notes how "unused" he is to travelling "normally" through time,[37] or asks "is this how time normally passes? Quite slowly...and in the right order?"[38] Similarly, for *Doctor Who,* as for any fictional text, the very essence of traditional narrative structure is the split between the events as they are, and the representation of those events as they occur on the screen (or in the text). In the previous chapter I discussed the difference between the story and the discourse, and now I want to turn back to this distinction and analyze a further dimension of the role that time travel plays in depicting narrative temporality.[39]

Although he does not specifically mention the pairing of the story and the discourse, this aspect of narratology haunts Jason Mittell's analysis of complex narrative structures on television.[40] To extend Mittell's analysis from fictional television to non-fictional history, however, there exists an interesting relationship between the story as a reconstruction of narrative and history as a reconstruction of the past. For example, when reconstructing the (non-televised) events that make up the story of the often-mentioned but rarely seen Dharma Initiative on the television show *Lost,* one must piece together hints and clues dropped into the discourse of the show by the producers; for instance, the video with background on the Swan station in the episode "Orientation." We may piece together the story from moments in the episodic discourse, or we may use a transmediated

approach, putting ancillary products into a cohesive narrative whole.[41] Audiences use this strategy to compile complete backstories of characters; for example, figuring out what really happened to *Boomtown's* Fearless in Iraq. Or, to turn back to *Doctor Who*, the show often reveals discursive "holes" in the narrative—an "endlessly deferred narrative," as Matt Hills calls it—that provide entry points for understanding and developing an underlying narrative story to the show.[42] One such discursive hole is the Doctor's age, which increases at a rate faster than that of the audience's viewing of the show: for example, in a serial broadcast in 1978 the Doctor's age is given as between 756 and 759.[43] In an episode broadcast in 2010, 32 years later, the Doctor states that his age is 907.[44] The complexity of the *Doctor Who* narrative makes it such that the "missing" 148 years of the Doctor's life *exist* but are *unseen*, a part of the story.[45] Indeed, one key role that fans play in establishing connective links in the show's mythology is to fill in the gaps in the discourse.

The understanding of history makes a similar demand on cultural participants as does narrative on television. Most events of history are unseen, and contemporary historians must reconstruct and place them into a narrative context. The fact that I can mention the Battle of Culloden and others can understand to what I refer, for example, means that we have all reconstructed this battle in our own minds, in our own ways. Each of our understandings may be different from each other, and we recognize that our personal reconstructions are not similar to what "actually" happened, but those mental representations *become* real to use. Yet, this memory itself is shaped by the multitude of cultural codes that go into the collective memory of the Battle of Culloden, including *Doctor Who*. As Confino describes, "the construction of the past, through a process invention and appropriation, affect[s] the relationship of power within society."[46] History has already become narrativized, as we cultural participants reconstruct a subjective understanding of our common cultural context. Nora argues that the concept of memory is subsuming history so that our collective memories overshadow our understanding of any historical event.[47] The consequences of this

historical reconstruction are immense: if our understanding of history is necessarily always re-created, then history itself can be changed.[48] In a more Whovian sense, history is always under construction: the first few seasons of the show featured a mix of science fiction serials and historical dramas. Although "*Doctor Who*'s portrayal of history was to be neither nostalgic nor reverential," it was always constructed from a Euro-British standpoint.[49] In his analysis of the early days of the program, Kilburn argues that the show's creators wanted to portray the "ordinariness" of history, but due to the demands of programming, had to make them artificially suspenseful, exciting, and action-packed.[50] Yet, even within this rigid entertainment structure, all *Doctor Who* historicals had to follow the "latest historical and scientific research," although each historical story also dramatized contemporary issues.[51] Early *Doctor Who* "fills the gaps" of *history's* hyperdiegesis.

In the ninth season episode "Day of the Daleks," for example, the Doctor and Jo must stop a band of guerillas from coming back in time from the 22nd century and changing Earth's history (at the present moment). Yet, in order to do that, the Doctor has to *allow* the Daleks to intervene in Earth's history—for that's the way it has "always" happened. Contrast this version of time travel with the first season's "The Aztecs," in which the Doctor forcibly declares that no one should change the course of history. It seems as though intervention in history is important sometimes, but not other times. As Hills writes, "if history were always fixed then there would be nothing [narratively] at stake."[52] For the Doctor, it is important that history remains on track. He explains the temporal situation to Jo at the end of "The Day of the Daleks":

> somehow the Daleks managed to pervert the course of history so they could conquer the Earth. The guerillas tried to change things back, but because they were part of history, their intervention just repeated the pattern. I was able to intervene and put history back on its proper tracks.[53]

In the thirteenth season's "Pyramids of Mars," the Doctor and Sarah try to prevent Sutekh, a hostile alien, from destroying the Earth in 1911. When Sarah mentions that Earth *wasn't* destroyed and that in her time it is quite verdant, the Doctor travels with her to an alternate "present" of her 1980, a present that emerges from the particular point in time when they *don't* stop Sutekh. The planet is a desolate wasteland. Although the Classic *Who* cautions against changing history, this is the only time it actually shows us that history, or at least a fictionalized and narrativized sense of history, *can* be changed.

The difference between these older historical episodes of *Doctor Who* and the New *Who* series lies in the way each constructs history. The nihilistic conception of history in "The Day of the Daleks," and "Pyramids of Mars" lies in direct contradiction with the New *Who* series of temporal discontinuities, which tend to emphasize the generative power of action.[54] Yet, in *Doctor Who,* as indeed in any historical reconstruction, history is a construction that has already become narrativized. We, as cultural participants, reconstruct a subjective understanding of our common cultural context. Eco references this idea in his semiotic analysis of the "reality" of fictional characters when he writes:

> not only fictional assertions but also the historical ones are *de dicto*...in other words, except for judgments depending on my direct experience (of the kind *it's raining*), all the judgments I can make on the grounds of my culture experience...are based on textual information and, even though they seem to express *de facto* truth, they are merely *de dicto.*[55]

The consequences of this historical reconstruction are immense: if our understanding of history is necessarily always re-created, then *history itself* can be changed. Writers like George Orwell understood this quite well: protagonist Winston in *1984* works as a historical revisionist, literally re-writing the history of Oceania to coincide with The Party's official version of history (the war between Oceania and Eurasia has always happened, even when Oceania is at war with Eastasia). To be able to re-write not only the facts but also the

interpretation of those facts lends great power to those with the pen—or keyboard. History can be updated and files lost, as Kindle owners found out in 2009 when Amazon (in a delicious irony) deleted electronic versions of George Orwell's *1984* and *Animal Farm* from users' Kindles. The digital information was erased, users' money was electronically returned to their bank accounts, and media history was erased, instantly.[56] *1984* indeed.

As a representation of that re-writability, wikis become a technology of historical uncertainty—that is, they symbolize the mutability of historical narrative. As a literal representation of the melding of narrative and history, *tardis.wikia* presents the historical existence of *Doctor Who*'s events as rewritable nuggets of narrative information. The *Doctor Who* narrative (re)presented on *tardis.wikia* becomes preserved as a historical story within the framework of the wiki/discourse itself. By examining more closely the way narrative production functions on *tardis.wikia,* we can develop a framework for understanding the presence of history itself within a wiki framework.

The Historical Codes on *Doctor Who* and *Tardis.wikia*

As present in and around the text of *Doctor Who,* four historical codes illustrate the relationship between narrative and history. Due to the dense relationship between the narrative of time travel, the length of the show's run, and the inclusion of actual historical events within the text of the show, *Doctor Who* offers a unique mechanism to see these codes. Furthermore, these codes are also reflected in the way wikis create and maintain a relation between narrative and history. In this section, I examine the Meta Code, the Diegetic Code, the External Code and the Future Code of *Doctor Who,* and illustrate how they compare and emerge on *tardis.wikia* as exemplars of the talk page, the history page and the rewritability and citation styles of wikis.

Meta Code

I use the term Meta Code to refer to any discourse that intra-textually references its own production. The Meta Code on *Doctor Who* develops because the nature of the television show allows for multiple authorship and collaborations through time. It is common for television shows like *Doctor Who* to have many people writing episodes, constructing backstories, illustrating key points, or adding to the narrative. By illustrating its own history within the narrative context of the television text, *Doctor Who* acknowledges the collaboration of the many people who worked on the show throughout the years; and serves as a metaphor for the way knowledge *as a collaborative product* is created by groups on wikis.

Doctor Who features multiple moments of self-reference. In the 1988 story "Remembrance of the Daleks," for example, the Doctor and Ace travel to the location and time that immediately preceded the events of *Doctor Who*'s first episode, "An Unearthly Child." That premiere episode took place on the same day it aired, 23 November 1963—that is, the diegesis was contemporary with the air date. "Remembrance of the Daleks," filmed as part of the show's 25th anniversary, focuses much of its narrative on a reflection of and comment about the *Doctor Who* series as a whole. But no allusion is more clear or more specific than when Ace is watching television and the announcer intones: "This is BBC television, the time is quarter past five and Saturday viewing continues with an adventure in the new science fiction series *Doct-*" [and then the scene cuts away].[57] By mentioning its own premiere, *Doctor Who* not only self-consciously describes its own creation, but also self-referentially circles back to articulate a connection to a particular moment in its diegetic history.

A more recent episode tells a subtler example. In "Human Nature" the Doctor has—temporarily—given up his identity and replaced it with a false one. He believes his name is John Smith and that he is a school teacher in 1913. As part of the false memory construction, the Doctor also believes that his parents (or, rather, John Smith's parents) are named Sydney and Verity. It is difficult to believe that these character names were chosen at random, as they reference

the original creator and producer of *Doctor Who,* Sydney Newman and Verity Lambert.

Meta Codes on *Doctor Who* appear as well in episodes that premiere on the date depicted in that episode. For example, the finale of the 2010 series revolved around the wedding of Amy Pond, which took place on 26 June 2010—the same day the episode aired. These meta-moments become self-reflexive comments on the time period of the viewing, taking present and turning it into mediated history.

In these Meta Codes, the information gleaned from watching *Doctor Who* comes to the viewer in two areas that also allow an understanding of wikis as texts. The first is the narrative information of the show itself, which increases with each additional unit of information given—that is, by telling viewers that the Doctor has landed on 26 June 2010 to attend Amy and Rory's wedding, *Doctor Who* has created a piece of narrative information that helps viewers to construct a complete view of the character, of the story and of the *Doctor Who* universe (a view that Parkin has used to write a revised history of the world according to *Doctor Who*).[58] The second aspect of information that arrives through the Meta Code is information about the development and creation of the television show *Doctor Who* as a television show—that is, so-called "behind-the-scenes" information that reveals how the show developed as a program from 1963 onwards.

It is this second type of information that most directly references and reflects the peculiar nature of wikis as both knowledge archives and as meta-comments on those archives, highlighting the use of Meta Codes in the writing of wikis. As Konieczny shows, dialogue about the construction of a wiki page can be as intense and as lengthy as the page itself: this dialogue is crucial to the success of Wikipedia, as the wiki forms on "a social institution, a part-community, and a part-social movement."[59] As wiki articles grow and develop, pages emerge that discuss the content on the original article. These pages, as Bruns points out, include "increasingly specific and complex policies for content creation and community collaboration."[60] By referring to

the page it is about, a "talk" page becomes tautological: it is what it does.

As Stvilia, et al., have noted, this dual purposing of wikis is one of the unique aspects that make them particularly useful for researchers: to reflect on the construction of a wiki article as it is being written allows for a practical functionality of constant progress.[61] Further linking discussion with content, Bruns writes that "community discussion about the content creation process...contributes to the development of content as well as community structures."[62] The result of this dual purposing on wikis acknowledges alternate viewpoints and reflects the multi-various and complex nature of historio-narrative construction. By facilitating group discussion *about* a topic, wikis illustrate that no topic has a singular viewpoint. And as *tardis.wikia* references *Doctor Who* as a constructive text, it assures multiple readings. In many ways, this is similar to type of postmodern criticism defined by Naremore as a "mixture of familiar discourses ... [that] can't be identified completely with any of them."[63] For Naremore, the cultural politics of (in this case) film criticism became influenced by the mediation of film on television: critics began "viewing films in no historical order, regarding the classic cinema as something distant or dying."[64] History became outweighed by the sheer volume of texts, the polysemic availability of media.

More concretely, we can see this in the article/discussion about the episode "The Big Bang" on *tardis.wikia*. Like any article for an episode of *Doctor Who* on *tardis.wikia*, this one features an episode synopsis, a plot description, the cast and crew list, references to other episodes, story notes, trivia for the episode, continuity reference and other citation material. The different elements are written in precise language and with the same Neutral Point of View (NPOV) that characterizes entries on Wikipedia. For example, the "episode synopsis" element reads: "The Doctor is trapped in the Pandorica, the TARDIS has been destroyed and silence has fallen. The only hope for all reality is a little girl who still believes in stars."[65]

A tab at the top of the page leads to the "talk" page which contains a discussion thread that expounds upon various aspects of the main article. For example, under a section titled "Let's Keep Rumours Clean," a poster by the name Delton Menace writes (all *sic*):

> The description of what is inside the Pandorica fits the description of the Doctor [...] and he is shown being dragged to the Pandorica in a released picture most have seen by now. There has been a [p]opular rumour that Omega [is] returning, too, and some reliable spoiler people on Gallifrey Base (these are people who, in all honestly, know quite a way ahead, some see unreleased pictures earlier on, [etc]..) may not know if the rumour is true or not, but they mostly say it is possibly -- nothing even they know suggests it can't be.[66]

And this is but one of 49 separate sections of hundreds entries, each of which has thousands of user-generated words written about it, more than four times as much as the page about which it talks. Because the "talk" page is not only connected to, but actually is a part of the article page, it becomes a tangible part of the wiki text. Any references become intra-textually articulated, and much like the meta-moments on *Doctor Who,* factor into any larger understanding of the text.

Diegetic Code

A similar type of connection between the television series and the wiki exists with what I call the Diegetic Code. The Diegetic Code articulates moments that reference previous episodes, events or characters as aspects of the particular narrativized history within the show's texts. Mittell has alluded to a similar code with his "Diegetic Retelling," when characters mention in-text important details for the audience to remember.[67] For example, in Chapter Two I mentioned the scene in *Doctor Who* when the Doctor rises up on a rock in the middle of Stonehenge in 2010's "The Pandorica Opens" and yells to the spacecraft circling the planet to "Remember every black day I ever stopped you, and then do the smart thing. Let somebody else try first." The dialogue not only allows his enemies to recall the events that have previously led to their defeat, but he concurrently speaks to the viewers and memorializes previous episodes, previous components of a diegetic history, that have undergirded the narrative of *Doctor Who*.[68]

In fact, it is a rare *Doctor Who* episode that does not reference, directly or obliquely, the events in the show's timeline that have previously aired. Such allusions are common and help to structure a show that has a 50-year history. The references can be minor—exclamations of "The Daleks are my most formidable enemy"—or can be major, as in "The Big Bang," when the Doctor is thrust backwards through his own timeline and actually witnesses or interacts with his previous self or companions. One such palpable event actually figures into the time-twisty narrative arc of the 2010 series: as the Doctor hurtles backwards through his own timeline, he arrives in a scene the audience has previously watched in "Flesh and Stone." Amy Pond, temporarily blinded, has just said good-bye to the Doctor when he comes back and monologues a speech about her memory of him. As I described in Chapter Three, during an initial viewing of "Flesh and Stone," such a monologue might seem slightly out of place: his insistence and his vigor (not to mention his outfit) do not seem to fit the scene. The inconsistency is later explained in "The Big Bang," when it is revealed that the audience actually saw a *future* Doctor in "Flesh and Stone" coming back in time. To watch "The Big Bang" is to be enveloped within a diegetic history that forces the audience to rewrite entire sections of the previously-viewed narrative.

The Diegetic Code represents a way of looking at the story canon as something akin to "fictional fact."[69] That is, although *Doctor Who* is fictional, the amount of narrative history and the adherence to a canon of elements that remains constant from one iteration to the next lends credence to these events, and gives them the appearance of factuality in the fictional space. This fact/fiction amalgam is one way, according to Gwenllian-Jones, that fans become enamored of a media text and treat it as an emotionally relevant portion of their lives.[70] Jane Espenson mirrors this assertion, arguing that the writing of a cult text involves the specific choice to make a fictionalized world seem as real as possible.[71] As a specific example of a Diegetic Code, the episode "The Next Doctor" features a man who has been implanted with all of the Doctor's memories, and thus thinks that he *is* the Doctor. In the dialogue between the two men, much of the in-show history of the character of the Doctor is revealed or recounted, and functions as a form of diegetic backstory for the story.

Along the same lines, a connection exists between the television show's narrative history and the narrativized history on wikis as symbolized by the "history" page. Every wiki entry has a "history" tab, which allows users to read through the revisions of the wiki in order to see which changes have been enacted. Just as the talk page on the wiki functions as a glimpse into a present-tense intra-textual discussion of the wiki, so too does the history page function as a way to look at the evolving and changing past "facts" of the wiki. The history page is one of the most important on a wiki, as it is a visual reminder that, as Roy Rosenzweig shows, wikis are a form both of collaborative history and of historical events.[72] Every change on a wiki is meticulously recorded and noted, and can be undone by anyone with a delete key. Our memories become shaped by what we read as history—and problematically, as Nora discusses, "memory and history...appear now to be in fundamental opposition."[73] Memory, like a wiki, is constantly evolving, shaping through cultural use. History, however, has today turned into a "problematic and incomplete" reconstruction of what memory visualizes. The wiki plays with and reverses this dynamic, creating history-as-memory and writes both as narrative.

Just as the Diegetic Code at play in *Doctor Who* is a continual reminder of the vast network of narrative history of the show, so too is the history page of a wiki a continual reminder of the vast network of historical narrativizing that has characterized the wiki article. It is not just that each page can be authored by many people, but that the history of that authorship can be recalled and analyzed at will. For example, on the aforementioned "The Big Bang" page, there are over 500 different versions, each recorded on the history page. The first "The Big Bang" page was started at 16:09 on 19 May 2010 by the user Kerange, and existed solely as the following line: "The Big Bang is the second part of The Pandorica Opens and the final episode of Series 5.

It's name was released on the official website." This was followed five minutes later by the user Solar Dragon, who instigated a "delete" command. In other words, Solar Dragon wanted to erase the page. Three minutes after that, the user Michael Downey added an enormous amount of information for the as-of-yet unaired episode, including possible spoilers, cast list, a synopsis, and DVD information. In less than ten minutes, a history had been created.[74] The history page is a reflection of the diegesis of the wiki. Like a continuing reminder, the wiki history becomes a diegetic retelling of the "story" of the article through time.

External Code

The third code of history exemplified both by *Doctor Who* and by wikis is what I term the External Code. In this code, historical events that have actually occurred in humanity's past are represented in *Doctor Who* as events that are either witnessed by, or in some cases, created by the Doctor. These events may sometimes be a part of episodes as backdrops, as historical context while the plot of the episode focuses on something else, or they can sometimes feature as the main plot line. Importantly, these events are not always accurate or reliable. To filter events through the lens of *Doctor Who* means necessarily appropriating history—the beliefs, ideas, experiences, events, movements and moments of the past—as ideologically determined moments. That is, every historical moment in *Doctor Who* must be seen as tinged with present-tense meaning. Cook, although writing here about the cinema, describes how the representation of history onscreen contributes more to an understanding of the present than it does to that historical moment. [75] For example, any reconstruction of the past "is a matter of cut and paste, of cutting bits and pieces of meaning," and thus reflects a contemporary attitude towards the subject matter.[76] In the same way, similar External Codes are present on wikis, as wikis also represent historical situations with varying degrees of reliability.

Recent examples of the External Code on *Doctor Who* include "Tooth and Claw," in which the Doctor and Rose travel to visit Queen

Victoria, "The Shakespeare Code," in which the Doctor and Martha travel to 1599 and meet Shakespeare, and the aforementioned "Fires of Pompeii." The effect of this melding of history and narrative in the fictional canon of *Doctor Who* is that historical events actually become appropriated by the narrative of the show: it is no longer the eruption of Vesuvius, but rather the Doctor's version of the eruption of Vesuvius. Indeed, *Doctor Who* has, since its premiere, focused on Earth's history (particularly its Western Anglophone history), and has represented many different historical events within its narrative, including the Crusades, World War I, and even the Trojan War. For example, in the episode "Victory of the Daleks," the Doctor travels to WWII-era England during the Blitz and finds that his arch-nemesis, the emotion-less Daleks, are attempting to change Earth history. Like many battles between these two old foes, the Daleks call the Doctor's emotional core—his empathy—his greatest weakness, yet he consistently demonstrates that it is his greatest strength. Throughout the episode, the emotionless Daleks and the emotional Doctor are pitted against each other, with the backdrop of WWII and Winston Churchill serving as foils to this conflict. As the show has ideologically linked the Daleks with the Nazis throughout its history, it's not too much of a stretch to see the depiction of the "emotionalism" of Winston Churchill combatting the "emotionless" Nazi soldiers as bolstering the British sense of national pride.[77]

The appropriation of history is also at the heart of the functioning of wikis, as these online repositories of historical "truth" attempt to record history through collaborative authorship. Yet, as Bruns points out, it is not reality that is recorded, but rather "(interpretations of) reality."[78] History as we know it on a wiki becomes merely one possibility within a multitude of other possibilities, each presented (or, rather, re-presented) with little to no historical authority, each updated instantaneously, changing "history" with each iteration. The power of the wiki lies in the fact that anyone can contribute. The thought is, as more people contribute, the greater the breadth of knowledge that becomes available.[79] Yet, as more people contribute, more versions of history, each formed from the socio-cultural background of the contributor, "work actively to create another reality"—a reality shaped by multiple, updatable perspectives on history.[80] And indeed, in reality what is written is always constrained by ideological and social forces. History, as the saying goes, is always written by the victors: and mediated history merely makes this literal, as history on wikis can only be written by those with access to a computer. According to data published by Nielsen Online, only 30.2% of the world's population use the Internet, with the fewest percentage of users residing in Africa, Asia, and the Middle East.[81] As a form of this "collective intelligence" of humanity, wikis don't so much represent the collation of humanity's knowledge as much as they do the Western, Euro-centric view of knowledge and history. When one can rewrite anything, the notion of authoritative truth disappears and a culturally specific truth appears.

For example, the article about "The Big Bang" on *tardis.wikia* has hundreds of contributors, each adding text to the article. Some users, like Michael Downey, have made major changes, writing vast sections of text. Other users may make only one or two minor edits, correcting grammar or spelling, or adding only a few sentences. In the first three days of the article's creation, for example, over 20 separate contributors made over 50 changes to the site: adding cast members, trivia items, ploy synopses, and references to secondary texts. The users continue to grow the article, and the history of *Doctor Who* becomes recorded instantaneously.

Ultimately, one way to understand the complex play with historical truth that wikis enact can come from the way television programs construct the nature of history within the fictional boundaries of the show. How, for instance, does a fictional account of a factual event factor into our everyday interpretation of that event? As represented in fiction, history as a concept is not only popularized, but is also problematically accepted as truth.[82] In fictional programs, history often becomes subsumed under narrative necessity: events shortened, characters amalgamated, places changed. The Doctor's Culloden is not the actual Culloden, but then again, neither is the one in my mind, nor anyone's. These are generally accepted as necessary changes for fictional programs, in order that the narrative be exciting, be involving or be popular. Similarly, a different set of changes are enacted on wikis, as historical "truth" can be extrapolated to mean any number of versions through the inherent re-writability of the wiki technology. Wikis illustrate, in the way that fictional narratives portray, an inherent mutability in the historical truth of an event. Or, to put it another way, wikis illustrate how "narrative [as history] becomes a problem only when we wish to give to real events the form of story. It is because real events do not offer themselves as stories that their narrativization is so difficult."[83] Importantly, if historicity becomes fact, then understanding the process of construction becomes paramount.

Future Code

A fourth code that *Doctor Who* makes salient represents future events: that is, it recounts events that *have* previously happened in the future but that *have not* yet happened in the present. In Chapter Two I discussed the issue of flashforwards in television programming as a way of illustrating future events. But the Future Code of *Doctor Who* and of the wiki illustrate a more complex play with temporality. As a show that focuses on time travel as a narrative device, *Doctor Who* often deals with causal paradoxes, or the establishment of effects before the causes that have happened. In the last chapter I mentioned Rory's discovery of the future Doctor who had already escaped from the Pandorica while the present Doctor was still trapped in the Pandorica. Similarly, another example of this type of Future Code appears in the aforementioned "The Big Bang." The Doctor and his companions are running up the stairs of the National History Museum when a time travelling Doctor from the future appears before them and dies, whispering something to the present Doctor. The cause of his "death" is shown later in the episode, when a Dalek exterminates him.

Other Future Codes exist in episodes that focus on the future as if it were a historical past. The early "The Tenth Planet," for example, sees the Doctor and his companions landing on Earth in the future (a far distant 1986). Yet, for the Doctor, this is *his* history, and he lands knowing full well how the events are going to play out: "once confirmation arrives, he jots down a summary of what's going to happen next on a slip of paper" and tries to hand it to one of the protagonists.[84] Later series have focused on the future history of the planet Earth. When the Doctor and Rose land in New New York City on New Earth, for example, he explains that this is actually the 15th New York City. The "history" of the inhabitant of New Earth does not end there, however; for in an earlier episode, "The End of the World," the Doctor takes Rose to see the destruction of Earth in the year 5 Billion. In effect, Rose (and the audience) saw the destruction of Earth before she saw the rebirth of New Earth, jumbling events as well as causes and effects. Everything is "history" to the Doctor.

"Silver Nemesis" represents another type of Future Code, one where the result of a plot development is revealed before the plot's setup. In this episode, Ace sees a portrait of herself hanging in a hall in Windsor Castle in 1988, although she had not yet posed for it.[85] Other episodes feature similar time-shifts in causality. Amy Pond leaves the Doctor a note in the past that he tells her to write in the future ("The Lodger"). The Doctor later leaves Amy a note in a similar manner ("The Big Bang"). The two-part "Children in Need" specials from 2011 depict a Future Code in a Möbius-strip-like narrative, where the TARDIS lands inside the TARDIS. Amy enters/exits the TARDIS simultaneously but travels a few seconds in the future and is able to relate what happened in those few seconds, changing/creating the past. These relatively minor events form a causality paradox, where no beginning to the event exists. For example, in "The Big Bang," Amy finds the note, which is how the Doctor knows how and what he has to write—there is no start to this causation cycle.

A more complex play with temporality exists with the aforementioned River Song, who inhabits a divergent time frame from the Doctor's. River Song represents the tension of the irreconcilable with the inevitable, and with each appearance in the show, her time line becomes increasingly complex. At times she appears to know nothing about what adventures have come (chronologically) previous to her story, while other times she is much more knowledgeable than is the Doctor. As both characters are time travelers, their relationship is a jumbled mix of temporal shifts. The complex interplay of time/temporality and narrative within the time travel structure of *Doctor Who* hearkens to Gendler's analysis of time travel within the film *Primer*: just as the film plays with time, so too *Doctor Who* creates "a fractured experience of time that is potentially analogous to the altering of reality carried out by the characters in the narrative."[86] When we first see River Song in "Silence in the Library," she is completely out of synch with the Doctor, as the following conversation elaborates:

> River Song: Thanks.
> Doctor: For what?
> River: The usual. For coming when I call.
> Doctor: Oh, that was you?
> River: You're doing a very good job acting like you don't know me. I'm assuming there's a reason.
> Doctor: Oh, a fairly good one actually.
> River: Okay, should we do diaries then? Where are we this time? Going by your face I'd say it's early days for you, yes? So, um... Crash of the Byzantium. Have we done that yet? Obviously ringing no bells. Alright, um. Picnic at Asgard, Have we done Asgard yet? Obviously not. Blimey, very early days then. Oo! Life with a time traveler, never knew it could be such hard work. Um... Look at you. You're young.
> Doctor: I'm really not, you know.
> River: Oh but you are. Your eyes! You're younger than I've ever seen you.
> Doctor: You've seen me before then?
> River: Doctor, please tell me you know who I am.
> Doctor: Who are you?[87]

As the dialogue suggests, this is the first time the Doctor (and, hence, the audience) has met River Song, but for River it is the last time she will ever see him, having traveled in time with the Doctor previously. Her future is written already (in the following series we do see River Song and the Doctor at the crash of the Byzantium), but his has yet to be (literally) written for the audience.

On wikis, a form of Future Code also exists, but in a slightly different form than with the anti-linear narrative plot of *Doctor Who*. Wikis have another type of Future Code, one built in the system of reputation and reliability of the articles themselves. As articles, wikis should include sources on all facts and cites for all information. *Tardis.wikia*, for example, follows in the style of Wikipedia to include footnotes for all facts in the article and asserts the canon of the show with hypertextual links to other articles. For instance, the *tardis.wikia* article on "Series 5" of *Doctor Who* includes fifteen footnotes to various articles, blogs, interviews and documentaries that have discussed the series, and scores of hyperlinks to other episode pages.

Yet, this citation system can be problematic as well, as the *assertion* of reliability rests on the inherent *impunity* of those citations. If a citation exists—even if it is incorrect or inauthentic—it can be attributed and used on wikis with authority. The problem with this reliability is that some wiki entries can be created based solely on the fact that *wiki entries exist based on that fact*. Using the same kind of circular logic that confused the authorship of the note from Amy to the Doctor, Jason Mittell offers an example of how his own Wikipedia entry was supported through reflexive citations to his own Wikipedia article.[88] Specifically, in a piece for *Le Monde Diplomatique*, Mittell defended the editability of Wikipedia against its critics despite the fact his own entry was scheduled for deletion.[89] But because of his piece's discussion of his Wikipedia entry's deletion, his Wikipedia entry actually thrived—as he writes, "the article about my Wikipedia entry has become a source to justify my inclusion on Wikipedia."[90] The future of Mittell's entry was assured by his own past challenge of it.

Mittell's experiences with Wikipedia echo with the Future Code on *Doctor Who*. The causality of an event is constantly in question when time travelling, because the effect can happen before the cause. Although such complex temporal mechanics may be confusing, understanding them may help elucidate the very real issue of reliability of citation on wikis. To debate the inclusion of specific events as history on a wiki means necessarily that someone thinks those events are important enough to include—a self-fulfilling prophecy. But by paying attention to issues of temporal uncertainty in popular culture we can begin to unravel the complexities of attribution on wikis, and the importance of authority in the age of historicity.

Conclusion

As Don Tapscott and Anthony Williams state, Wikipedia's "openness is also the reason why it is constantly growing, adding new entries, covering new niches and always reviewing and updating facts. It taps an almost infinite wealth of talent, energy, and insight."[91] Wikipedia is not alone: wikis on the whole adapt to the changing technological landscape with amazing versatility. Yet, some inherent aspects of the technology do remain constant: wikis are always editable, are updatable and reflect a continually changing sense of what is important *now*, at *this* moment in time, in *this* particular cultural milieu.

Transgenic media enact a flux within our cultural life. I believe that wikis represent a novel form of historical production. At the same time, however, wikis fall prey to the same types of ideological issues present in any form of cultural production. Only by understanding the limitations as well as the possibilities can users come to a full understanding of the capabilities of wikis as new media repositories of cultural history. History, in this cultural sense, is inherently communal. We share history in order to be part of communities, and we use history as a way to barter hierarchies in social status. To control history—to literally write history in our own words—is to represent one particular version of an event, as any history is necessarily a narrativized account of an actual moment in time. Wikis become a way of representing the mutable nature of history.

This fact, that knowledge is inherently narrativized and mutable, can be downright scary. The Orwellian nightmare of a rewritable history is antithetical to a democratic society. Yet the wiki, the most democratic of all new media technologies, allows this very mutability. To engage with transgenic technology, however, takes knowledge and experience. Often, we look to the familiar to scout out relationships to the new. In this chapter, I have demonstrated how traditional media texts, like *Doctor Who*, can act as heuristics for understanding new media technologies like wikis. As an epitome of this type of heuristic, the wiki *tardis.wikia* represents a joining of narrative knowledge production and wiki historical knowledge production. Through the four historical codes discussed in this article—the Meta Code, the Diegetic Code, the External Code and the Future Code—we can understand some of the more complex characteristics, including talk and history pages, and the rewritability and credibility of wikis.

If, as Halavais states, "knowledge is a vital part of democratic self-rule," then it is crucial to maintain a strict understanding of the technologies of knowledge production for a healthy society.[92] To look at the nature of knowledge, of history, of narrative, or of the interaction between all three is to focus on central tenets of our society. If, as the Doctor forewarns, "time is being rewritten," it is imperative to learn not just who does the rewriting, but also how that rewriting functions in a socio-cultural context. Although *Doctor Who* and *tardis.wikia* may be fictional texts, the characteristics they describe are applicable in many realms of cultural creations.

Therefore, as we have seen, transgenic media as represented through time travel narratives encompass two different ways of interpreting the relationship between narrative and history. Similarly, this representation articulates both the permanence as well as the fluctuations within our quotidian media usage. In the next chapter, I hope to reconcile these two opposing views by examining the network as a key metaphor in our contemporary media environment, both online and on television.

Notes

1 See both Hanley and Nahin.

2 Jean Baudrillard, *The Illusion of the End*, trans. Chris Turner (Stanford, CA: Stanford University Press, 1992), 26.

3 Alex Eichler, "Meet Keith Urbahn, First to Tweet News of Bin Laden," *The Atlantic Wire*, 04 May 2011, http://www.theatlanticwire.com/national/2011/05/meet-keith-urbahn-first-tweet-news-bin-laden/37331 (accessed 29 May 29, 2011).

4 Zizi Papacharissi, *A Private Sphere: Democracy in the Digital Age* (Cambridge, UK: Polity Press, 2010); David Beer and Roger Burrows, "Sociology and, of and in Web 2.0: Some Initial Considerations," *Sociological Research Online* 12, no. 5 (2007): http://www.socresonline.org.uk/12/5/17.html (accessed 15 Aug 2011).

5 Steven Moffat, writer, "Blink," *Doctor Who* (London: BBC, 2007).

6 For ease of reference, I will refer to this site as *tardis.wikia*.

7 See Paul Booth, "Narractivity and the Narrative Database: Media-Based Wikis as Interactive Fan Fiction," *Narrative Inquiry* 19, no. 2 (2009); Mittell, "Sites of Participation."

8 My analysis here mirrors a similar kind of analysis in Karin Littau, "Media, Mythology and Morphogenesis: *Aliens*™," *Convergence: The International Journal of Research into New Media Technologies* 17, no. 1 (2011), who examines the representation of the *Alien* franchise across media.

9 See Raymond Williams, *Culture and Society, 1780–1950* (New York: Columbia University Press, 1958).

10 Hayden White, *The Content of the Form: Narrative Discourse and Historical Representation* (Baltimore, MD: The Johns Hopkins University Press, 1990), 4.

11 Hayden White, *Metahistory: The Historical Imagination in Nineteenth-Century Europe* (Baltimore, MD: The Johns Hopkins University Press, 1973), 318–85.

12 Jameson, 6.

13 Caldwell, *Televisuality*, 164–65.

14 Ibid., 165.

15 Robert A. Rosenstone, *History in Images/History in Words: Reflections on the Possibility of Really Putting History onto Film* (Boston, MA: Harvard University Press, 1995), 4.

16 Eckart Voigts-Virchow, "History: The Sitcom, England: The Theme Park—*Blackadder*'s Retrovisions as Historiographic Meta-TV," in *Narrative Strategies in Television Series*, ed. Gaby Allrath and Marion Gymnich (Houndmills, UK: Palgrave Macmillan, 2006), 211–12, 217.

17 Ibid., 212, 217.

18 See Vivian Sobchack, *The Persistence of History: Cinema, Television and the Modern Event* (New York: Routledge, 1995).

19 Booth, "Narractivity."

20 Booth, *Digital Fandom*, 106.

21 Bruns, 103.

22 Jonathan Crary, "Spectacle, Attention, Counter-Memory," *October* 50 (1989): 98.

23 See Matt Hills, "Absent Epic, Implied Story Arcs, and Variation on a Narrative Theme: Doctor Who (2005–2008) as Cult/Mainstream Television," in *Third Person: Authoring and Exploring Vast Narratives*, ed. Pat Harrigan and Noah Wardrip-Fruin (Cambridge, MA: The MIT Press, 2009), 333–43; John Tulloch and Manuel Alvarado, *Doctor Who: The Unfolding Text* (New York: St. Martins Press, 1984); Hills, *Triumph*.

24 Tulloch and Alvarado, 179.

25 Guy Debord, *The Society of the Spectacle* (1967), trans. D. Nicholson-Smith (Detroit: Black & Red, 1983), 158.

26 Crary, 106.

27 Debord, 155.

28 The language is directly copied from the BBC summary of the episode before it premiered on 03 Apr 2010, "The Eleventh Hour," *Tardis.wikia*, http://tardis.wikia.com/wiki/ The_Eleventh_Hour#Synopsis (accessed 02 Apr 2010, 2100 GMT).

29 Ibid., (accessed 14 Apr 2011).

30 White, *Content of the Form*, 1.

31 Stephen Robertson, "What's Wrong with Online Readings? Text, Hypertext, and the History Web," *The History Teacher* 39, no. 4 (2006): 445–46. http://www.historycooperative.org/journals/ht/39.4/ robertson.html.

32 Ibid., 447.

33 Beer and Burrows, 3.1.

34 Ibid., 3.6.

35 Ibid., 3.7.

36 Adam Smith, writer, "Flesh and Stone," *Doctor Who* (London: BBC, 2010).

37 Moffat, "The Big Bang."

38 Richard Curtis, writer, "Vincent and the Doctor," *Doctor Who* (London: BBC, 2010).

39 Chatman, *Story and Discourse*; Genette, *Narrative Discourse*; Richard Walsh, "Fabula and Fictionality in Narrative Theory," *Style* 35, no. 4 (2001); Walsh, *Fictionality*; David Gorman, "Reconsidering Narrative Theory Rhetorically," *Poetics Today* 31 (2010).

40 Mittell, "Narrative Complexity."

41 See Lynnette Porter and David Lavery, Lost*'s Buried Treasures,* 3rd ed. (Naperville, IL: Sourcebooks, INC., 2010); also, Henrik Örnebring, "Alternate Reality Gaming and Convergence Culture: The Case of *Alias*," *International Journal of Cultural Studies* 10, no. 4 (2007).

42 Hills, *Fan Cultures*, 134; see also Christy Dena, "Emerging Participatory Culture Practices: Player-Created Tiers in Alternate Reality Games," *Convergence: The International Journal of Research into New Media Technologies* 14, no. 1 (2008).

43 Robert Holmes, writer, "The Ribos Operation," *Doctor Who* (London: BBC, 1978).

44 Smith, "Flesh and Stone."

45 Tat Wood, *About Time: The Unauthorized Guide to Doctor Who, 1985-1989, Seasons 22 to 26, The TV Movie* (Des Moines, IA: Mad Norwegian Press, 2007), 92.

46 Alon Confino, "Collective Memory and Cultural History: Problems of Method," *The American Historical Review* 102, no. 5 (1997): 1393.

47 Pierre Nora, ed. *Les Lieux de Mèmoire, Vol 7* (Paris: Les Frances, La Republique, Le Nation, 1992), cited in Jeffrey K. Olick, and Joyce Robbins, "Social Memory Studies: From 'Collective Memory' to the Historical Sociology of Mnemonic Practices," *Annual Review of Sociology* 24 (1998): 120–1.

48 Debord, 131–34.

49 Matthew Kilburn, "Bargains of Necessity? *Doctor Who,* Culloden and Fictionalising History at the BBC in the 1960s," in *Time and Relative Dissertations in Space: Critical Perspectives on Doctor Who,* ed. David Butler (Manchester, UK: Manchester University Press, 2007), 70–1.

50 Ibid., 71.

51 Ibid., 74–7.

52 Hills, *Triumph*, 95.

53 I have taken this quotation from the novelization of that episode, Terrance Dicks, *Day of the Daleks* (London: Target Books, 1974), 137.

54 Hills, *Triumph*, 87–89.

55 Eco, "On the Ontology," 86.

56 David Pogue, "Some E-Books Are More Equal Than Others," *New York Times*, 17 July 2009, http://pogue.blogs.nytimes.com/2009/07/17/some-e-books-are-more-equal-than-others/ (accessed 10 Apr 2011).

57 See Lance Parkin, *AHistory: An Unauthorized History of the Doctor Who Universe* (Des Moines, IA: Mad Norwegian Press, 2006).

58 Ibid.

59 Piotr Konieczny, "Governance, Organization, and Democracy on the Internet: The Iron Law and the Evolution of Wikipedia," *Sociological Forum* 24, no. 1 (2009): 166.

60 Bruns, 113.

61 Besiki Stvilia, Michael B. Twidale, Linda C. Smith, and Les Gasser, "Information Quality Work Organization in Wikipedia," *Journal of the American Society for Information Science and Technology* 59, no. 6 (2008).

62 Bruns, 109.

63 James Naremore, "Authorship and the Cultural Politics of Film Criticism," *Film Quarterly* 44, no. 1 (1990): 17.

64 Ibid., 18.

65 http://tardis.wikia.com/wiki/The_Big_Bang (accessed 18 June 2010).

66 Delton Menace, "Let's Keep Rumours Clean," *tardis.wikia*, 2010, http://tardis.wikia.com/wiki/Talk:The_Big_Bang (accessed 18 June 2010).

67 Mittell, "Previously On."

68 Moffat, "The Pandorica Opens."

69 A story canon is the complete backstory of a fictional text.

70 Sara Gwenllian-Jones, "Virtual Reality and Cult Television," in *Cult Television*, ed. Sara Gwenllian-Jones and Roberta E. Pearson (Minneapolis: University of Minnesota Press, 2004).

71 Espenson.

72 Roy Rosenzweig, "Can History Be Open Source? Wikipedia and the Future of the Past," *Journal of American History* 93 (2006).

73 Pierre Nora, "Between Memory and History: Les Lieux de Mèmoire," *Representations* 26 (1989): 8.

74 http://tardis.wikia.com/index.php?title=The_Big_Bang&dir=prev&action =history.

75 Pam Cook, *Screening the Past: Memory and Nostalgia in Cinema* (New York: Routledge, 2005), 218.

76 Ibid., 221.

77 Tat Wood and Lawrence Miles, *About Time: The Unauthorized Guide to Doctor Who, 1963–1966, Seasons 1 to 3* (Des Moines, IA: Mad Norwegian Press, 2006): 36.

78 Bruns, 119.

79 Pierre Lévy, *Collective Intelligence: Mankind's Emerging World in Cyberspace* (1994), trans. Robert Bononno (New York: Perseus Pub, 1997); Jenkins, *Convergence.*

80 Robert Hassan, *The Information Society* (Cambridge, UK: Polity Press, 2008), 150.

81 Internet World Stats: Usage and Population Statistics, "Internet Users in the World," (Mar 2011), http://www.internetworldstats.com/stats.html (accessed 16 Aug 2011).

82 See, for example, Rosenstone; Robert Toplin, *History by Hollywood: The Use and Abuse of the American Past* (Champaign: University of Illinois Press, 1996).

83 White, *Content of the Form*, 4.

84 Tat Wood and Lawrence Mills, *About Time: The Unauthorized Guide to Doctor Who 1966-1969, Seasons 4 to 6* (Des Moines, IA: Mad Norwegian Press, 2006), 18.

85 Available on a deleted scene on the DVD.

86 Jason Gendler, "*Primer*: The Perils and Paradoxes of Restricted Time Travel Narration," *Nebula* 3, no. 4, http://www.iiav.nl/ezines/web/Nebula/2009/No2/Gendler.pdf (accessed June 2010): 142–43.

87 Steven Moffat, writer, "Silence in the Library," *Doctor Who* (London: BBC, 2008).

88 Jason Mittell, "Trapped in a Web of Meta-Wikipedia," *Just TV*, 06 May 2009, http://justtv.wordpress.com/2009/05/06/trapped-in-a-web-of-meta-wikipedia/ (accessed 06 May 2009).

89 Mathieu O'Neil, "Wikipedia: Experts Are Us," *Le Monde Diplomatique English Edition*, 15 May 2009, http://mondediplo.com/2009/05/15wikipedia (accessed 01 June 2010): ¶10.

90 Mittell, "Trapped," ¶4.

91 Don Tapscott and Anthony Williams, *Wikinomics: How Mass Collaboration Changes Everything* (New York: Penguin, 2006), 75.

92 Alex Halavais, *Search Engine Society* (Cambridge, UK: Polity Press, 2009), 111.

Six

Temporal Displacement as (Social) Networking

If only we could both see from the other's perspective, as well as our own, at the same time. If we could do that, then we would have it all, the past and the future, fused, combined into one perspective, we would see the present moment, how it divides us, like mirror images around a temporal axis.

—Yu, How to Live Safely in a Science Fictional Universe, 226

As described, Yu doesn't just see things from one perspective. Through time travel, Yu mashes up multiple viewpoints to form a network of related positions. Yu's travels see him venture across multiple vantage points on his own life. The idea of the network pervades Yu's concept of time travel, for when people return to those events that are crucial to their lives, they can only look at them from the "outside looking in, from all sides."[1] That is, a time traveler is not allowed to have a first-person perspective on his/her life, but rather must only view events as an outside observer. Travelers see those events as shared moments experienced communally. They can't join in the celebratory birthday parties or the abject despair of funerals: they can only watch. But in doing so, the travelers see a variety of different perspectives. Through the reconstruction of these different points-of-view—of the first-person memory and the third-person time traveler view—the traveler creates a networked version of events, one that illuminates far better than either view can by itself.

This concept of the social network has also infiltrated recent discussions of technology, sociality, and televisuality. Social networkization has always been a component of fictional narratives, especially those that overflow with characters. One could easily argue that *The Iliad*'s large cast of characters and engaging temporality, well before the *Science Fictional Universe,* foreshadowed much complex television content. If television characters aren't reflecting on others, they're working with others, connecting with others, or engaging in dialogue with others.[2] The nature of this sociality, however, has changed on television in the past decade. Television emphasizes complex relationships between scores of characters and relies on the audience's inherent understanding of the network of relationships between characters.

Although today's character complexity is central to television narrative, character complexity can be seen in shows like *Hill Street Blues* (1981), which also relied on large casts. But these early shows were either relatively unique on television, or were niche (e.g., soap operas).[3] The soap opera arguably functions solely as a character-driven genre. Allen demonstrates how character relationships are crucial to understanding soaps: "the superficial meaning of a situation common to all soap operas is instantly overwhelmed by the deeper significance of that event in the specific character network in which it occurs."[4] Although characters have always been a central facet within television analysis, today's networked characters in shows like *Lost, Flashforward, Boomtown, The Wire, Breaking Bad* (2008), or *Heroes* have become commonplace. What Allen described as a unique feature of soaps ("These complex character networks in American daytime soap operas distinguish them even from their prime-time counterparts") is now becoming standardized across prime-time narrative.[5] Yet, for Mittell, character relationships are actually *backgrounded* in these complex shows:

> While certainly soap opera narration can be quite complex and requires a high degree of audience activity to parse out the web of

> relationships and backstory evoked at every plot turn, narratively complex programming typically foregrounds plot developments far more centrally than soaps, allowing relationship and character drama to emerge from plot development in an emphasis reversed from soap operas.[6]

I want to argue that looking at television through the lens of social networking foregrounds these character relationships in order to elucidate in greater detail audience/character connection.

Today's television can rely on an audience's knowledge of complex relationships because of the inherent networkability of transgenic media: as Johnson points out, "the process of acclimating to the new reality of [online] networked communication has had a salutary effect" on our viewing practices and on our everyday communication.[7] The more we work in an online environment and engage in networked communication, the more our "brain will grow more adept at tracking ...intersecting relationships" in other media.[8] For example, the television show *Lost* features over twenty-five "main" characters over the course of its six seasons, each one linked to the others in at least one seemingly coincidental way. The connections between the characters form a linked network of relations, a web of intersections. This connectivity is noted in much writing about *Lost*, from popular press articles in *Wired*, for example, which feature "maps" of the relationships, or from the DVD box sets themselves, which have special features that graphically link characters together.[9]

It is not my intention in this chapter to argue that an increased *cast* necessarily leads to (or from) a change in the nature of our relationships. Rather, I want to evaluate the way that the *network* seems to have become a metaphor for today's television content, articulated through our transgenic media use. I argue that two "network" metaphors function together on contemporary television. On the textual level, television characters form pseudo-social networks, resulting in a sort of fictionalized representation of a social network site, which engenders audience identification with the network by "placing" audiences within a network of similarly placed fictional "friends." On the structural level, many television narratives use temporal displacement as a networking device to create links between plot elements (nodes) that the audience must connect (network) in order to make sense of that narrative. This narrative framework situates the audience in a particular network of spatial connections. At both these levels, shows with a network at their center rely on an audience's mediated memories of characters, and their immersion in a mediated space. At the end of this chapter, I will have presented a new way of examining the convoluted textual relationship between television and transgenic media. By looking at the unique connections between viewer, character, and text, we can think above and beyond the paradox of the historical stability outlined in Chapter Four and the experience of historical mutability discussed in Chapter Five to link history, narrative and knowledge with a larger, cultural network structure.

The Networkization of Television: Narrative and Character

In this chapter, I'll be defining a "network" in a general sense, as a structure with an inherent *interwoven complexity*. In this structure, elements are related to other elements in more than one manner, and objects intersect with one another multiple times. The network is a mesh of interconnected nodes, presenting many paths through one terrain, and is a common metaphor we use (and have used in the past) for the hypertextual connections on the web.[10] Recent Internet/media research about the network argues that conceptualizing our society through the lens of the networked web can help elucidate some of the issues of connectivity that undergird our conception of contemporary culture.[11] The *idea* of the network in all its manifestations, from the social to the technologic, is:

> becoming the nervous system of our society, and we can expect this infrastructure to have more influence on our entire social and personal lives than did the construction of roads for the transportation of goods and people in the past.[12]

Manuel Castells, further, posits that the flow of information shapes our relationship with industries, communities, and our selves into a "network" of connections that describes contemporary social organization.[13] Our contemporary web use doesn't just link *objects,* but also links *users.* We link our Twitter account with our Facebook, which is also connected to Skype and Spotify. I can view not only my own musical history, but my friends' as well (as long as they've accepted the Facebook/Spotify connection). Further, it's not that we focus our *life* online; rather, our multiple *lives* are spread out through our connections online. My multiple profiles, linked together, make *me* (my self) a network as well; or, at least, my digital alter-ego becomes one. Through these networked connections, we undergird our own relationship with the media environment at large.

The rise of transgenic media has brought with it another avenue of meaning for the term "network," which can be used to link two opposing conceptualizations of media audiences. As Fiske and Hartley show, we often conceptualize the television audience as one of two extremes:

> In most discussions of television the audience is characterized … as the individual viewer, or as the many millions of viewers for any one programme. But the audience normally experiences itself in neither of these ways. Of course each viewer is aware both of his or her individuality and of his or her membership of a large group. But while actually watching, most people are part of a family audience.[14]

That is, as viewers, we don't recognize ourselves as solitary, nor do we necessarily recognize ourselves as part of an immense global group, but rather we see ourselves as individuals within more personal and social networks. Translated to the computer screen, popular culture representations of Internet use depicts similar conceptions—either the anti-social individual (the lone "nerd"), or the amorphous collective mass that contributes to Wikipedia or writes comments on YouTube (the faceless mob).

As Somers states, narrative is inherently about relationships.[15] For Somers, the key function of narrative is not to *transmit a plot,* but rather to *connect audience* to plot, to unite "an individual's understanding of self and the construction of identity [with] a degree of purpose and unity."[16] Taking Somers' analysis of narrative as ontologically linked to identity and applying that conception to the social network mode of television criticism, we can see a new conception of audience identification with character.

Specifically, Somers describes how narrative and identity are intimately and ontologically connected: "it is through narratives and narrativity that we constitute our social identities."[17] But in order to make this claim, she has to differentiate between different levels of narrative. Traditionally, critics have seen narrative as representational; that is, narrative was seen as a form that imposed structure on the "chaos of lived experience."[18] Narratives in this conception are descriptive, illustrating one particular view of an experience. Somers argues for shifting the conception of narrative to be more ontological in nature: that is, narratives become prescriptive as they "guide actions...people construct identities...by locating themselves or being located within a repertoire of emplotted stories."[19] And how do we locate ourselves within stories? By examining our "relationships (connected parts) embedded in time and space, constituted by causal emplotment."[20] What television criticism needs to do is apply these postmodern methodologies to the relationships between characters on screen, in order to examine audience/character connections.

Transgenic media have influenced our conception of our own spectatorship. What Fiske and Hartley examine—that whenever we watch media we experience it *both* collectively and individually—means that we experience media within a network of other viewers. Even when by ourselves, we have an understanding of the collective viewership of those in immediate *social proximity* to ourselves. By social proximity, I refer to the network of people that are close to us emotionally, even if they are far from us geographically. Online social and transgenic media have made these connections stronger and more realizable than ever before. Shirky illustrates the power of social proximity in his recounting of the story of how a network of people came together to help recover a stolen cell phone.[21] Thousands read the story of the phone on a website and offered assistance in a variety of ways. In terms of television, we all have our own network of social proximity whenever we watch: for example, my friend Brian got a text message after *24* (2001), and my uncle finds out my opinions about this week's *Doctor Who* over email, even though he lives half a globe from me. Even though I may watch *Glee* (2009) alone and I may intellectually know that it's being watched by millions of others, it's only the middle of these two extremes, my network of social proximity, that I think about—the dozen or so of my friends and colleagues with whom I will chat the next day. Social media has expanded the "proximity" of social proximity—thanks to Twitter, chatting with the world about television entertainment has become more common. To facilitate this interaction, producers have started creating "character" tweets (tweets sent from characters on the show) that also seem to attempt to foster a community.[22] When audiences desire this form of active identification, social media tools engage to provide it. Other social network sites have recently been created that play on our real-world connections to media: the site GetGlue, for instance, asks users to share with other users what they're being entertained by.

Social proximity relates to the networked mentality we know and use in our everyday lives. The network doesn't just link the individual and the collective, but enacts a number of other linkages across a range of media and cultural boundaries. Daly, although writing here about the cinema, describes these linkages:

> The electronic age with its emerging economies is blurring … contradiction[s], and similarly, I posit, this moment in culture is reflected in our current cinema. Increasingly, for computer and mobile users, existence is in some intermedial zone of work and leisure; the experience of moving images through computer and digital technologies is interactive, blurring the lines between producer and consumer, spectacle and spectator, representation and information, as embodied by mashups and crowdsourcing. To represent these new sets of relations between art, culture, work, and

> relations of power, [we] must move beyond vision to engage thought.[23]

Although Daly calls these relations between elements "blurred," I think it's more appropriate to see them as "intra-textual," that is, networked within a larger whole, where different characteristics interact within that whole. In his network theory of society, Benkler argues that our culture has recently made two shifts. First, we've moved from an economy centered on goods to an economy centered on information and the manipulation of symbols.[24] Second, we've moved from a linear mode of communication to a "communications environment built on cheap processes with high computation capabilities, interconnected in a pervasive network."[25] In both cases, these cultural shifts have created what he calls a "networked information economy," or one in which "decentralized individual action...plays a much greater role than it did, or could have, in the industrial information economy."[26] The networked public sphere "enables many more individuals to communicate their observations and their viewpoints to many others, and to do so in a way that cannot be controlled by media owners and is not as easily corruptible by money as were the mass media."[27] We are becoming a society of networking, where even more traditional communication spaces are shifting. Tyron shows how the availability of films on demand through services like Netflix and Vudu have created a "film networked culture," where audiences in small towns can receive and discuss any number of non-mainstream films.[28] Wikis, as we saw in the last chapter, exemplify this type of social proximity, amalgamating the role of the author and the user. Blogs and Tumblr foster interactivity through comments and reposting.[29] Tweets are retweeted. It's not that networking is new, per se, but rather that the *experience* of the network has become normal and thus standardized.

We can see this experience of the network on television, through the lens of temporal displacement. By linking together the web of unstable elements—character, memory, story, discourse—within complex narratives, temporal displacement functions in the same way as a network does. Furthermore, a play with temporality in complex narrative engenders a networked understanding of plot on television. For example, *Lost* presents a multi-linear narrative, in which a change to the past does not necessarily conflict with events in the present. Instead, to change events in the past will create *alternate* futures, new timelines, and concurrent linearities. In terms of character, *Lost*'s score of protagonists and antagonists also create a network of connections. Audiences have to keep track of personal histories, connections with other characters (both major and minor), and maintain a continued sense of why they might be important on the island. In this way, *Lost* follows a "multithreaded" narrative and set of characters.[30] To help audiences track these characters and this narrative, *Lost* uses temporal displacement to represent different timelines that take place simultaneously in the diegesis. For example, during the fifth season of *Lost*, some characters (Sawyer, Juliet and Miles) jump from timeline to timeline until they find their way to 1977, some characters (Jack, Kate, and Hurley) remain in the present (2007), and some adventures take place on the island in 2004. Audiences have to remember how the events in 1977 affect the events in 2004, and how those events affect 2007. The "flash-sideways" plots of season six detail events that seem to happen in a parallel world, one in which our characters never crashed on the island. A networked understanding of the complex narrative was required in order to make sense of character development, plot mechanics, and, ultimately, the "truth" of the show.

This networking function of contemporary television can be seen across complex television. For example, both the characters and the narrative of *Doctor Who* are inherently networked, as the Doctor and River Song intersect in a variety of ways throughout time. As I've described, River Song first meets the Doctor sometime in his future but last sees him during her first appearance on the show. The two characters meet half a dozen times across a couple of series, but for each of them the meeting takes place at a different point in their relationship. There is a similar "networkization" of character and narrative in *Flashforward*, which presents a sprawling cast of characters connected in both the past and the future, and whose connections drive the plot forward.

Although large, the number of networked connections on *Lost* is no more considerable than the number of connections we make everyday in our digital life. We connect the events of *Lost* much like users of transgenic media must "connect" with friends on Facebook, might "connect" a web narrative like *Lonelygirl 15* from disparate YouTube clips, or could "connect" research and citations on Wikipedia to explore the representation of knowledge online. The temporal displacement of *Lost*'s characters matches our own narratological use of Facebook, for example. Searching for our friend's images of the party last week, our wedding last year, or the missed connection with a pal from college—when we Facebook "stalk," we are traveling in time, connecting events from days, months, even years ago through our hypertextual connections.

To follow a complex television narrative network like *Lost*'s necessarily means being actively engaged with the material. As Daly describes, this type of engagement has become common, a "natural consequence of digital media. Computer and digital users have been trained by their immersion in digital culture to participate in/with what they consume."[31] In the next section I examine how social networking in our everyday transgenic media saliently mirrors the characteristics and specifics of networked sociality in television through sprawling casts of characters linked by various diegetic interconnections.

Character Networks as Social Networks

I am not the first to make the connection between social networks and *Lost*. In *Beyond the Box*, Sharon Ross examines how the fictional survivors of the plane crash "are in need of a social networking force," one that "operates much like its fans, who network with each other in attempts to unravel the narrative 'messiness' of *Lost*."[32] But she does not investigate the ways the *text* of *Lost* mirrors characteristics of transgenic media (or vice versa). Ross instead focuses on the television industry's navigation of "the [economic] opportunities and demands of new media."[33] Her excellent analysis of the television industry, in which she argues that the *Lost* narrative allows "different ... threads to connect" across media boundaries, "each ... forcing the participant to reassess the story," also touches upon issues of audience participation and networked narrative.[34] Yet, she does not describe the larger *textual* connection between the show and social networking, a phenomenon that Johnson describes as "a thick network of affiliations."[35]

Given the rise in popularity of online social networking during the six years that *Lost* was on the air, it should come as no surprise that there could be a connection between transgenic media and the textual and structural aesthetics of the program. MySpace's 100 millionth account was created in 2008, during *Lost*'s fourth season.[36] In 2009, 83% of 18–24-year-olds had a social networking profile.[37] In 2010, the last year of *Lost*, Facebook had over 500 million subscribers.[38] I don't want to make a spurious and hyperbolic argument that *Lost* specifically used social networking characteristics exclusively *because of* the popularity of MySpace and Facebook, but I do want to note that a meteoric rise in the population of social networkers parallels a similar networkization in the show: and while *Lost* is not the only show that features this type of network organization, it is the most popular and visible. The audience identification with the characters' island social networks is engendered by the metaphor of the online social network. Specifically, like any profile in an online social network, each character on *Lost* first presents a constructed version of his/her own identity, and then fits that identity into a larger social framework with the other characters. By recognizing the characters in a social network, audience members can then see themselves within that network as well. Social networks can have real-world implications, providing the "glue" between activities both on- and offline.[39]

Identity Construction

Each character on *Lost* seems to present a constructed version of the self, creating a personal face to the world. This type of identity presentation is "enabled by a performative palette that combines multimedia elements with cultural references, elements of play, denotative and connotative expression, and a variety of tools."[40] That is, as sociologist Erving Goffman illustrates, we constantly perform our identities, creating images and characters depending on the situation and audience with whom we engage.[41] In our society, "the character one performs and one's self are somewhat equated, and this self-as-character is usually seen as something housed within the body of its possessor," but it is a *performance*, geared for a specific effect.[42] That is, I am not the same person when I am in front of a classroom as I am when I am in front of a dinner party. The labels we give (or which are given to us) illustrate the different roles we play everyday: professor, son, viewer, husband. Online social networks externalize and make visible the process of persona creation as a "fluid abstraction" of the self.[43] Of course, this type of identity construction is not new; it is merely fostered and expanded using online social networks.[44]

On *Lost*, we can see this type of persona creation with each of the main characters, as each tries to create a new identity for him/her self on the island. Kate is a prime example of this, as early on in the series she tries to hide the fact that she was a criminal being escorted to LA under the supervision of a US Marshal. Once they crashed, she was able to construct a new identity, a new persona, for herself. Sawyer, too, creates a new identity for himself, changing his name from James Ford to "Sawyer" in order to create the persona of a con man, and branding himself with literature, glasses, and scruff. Charlie at first tries to hide his drug problem. Locke doesn't reveal his paralysis and takes on the persona of a hunter. Ben pretends to be Henry Gale when we first meet him. When Sawyer, Jin and Miles are stranded in the 1970s, each of them has to construct new personas in order to fit in with the DHARMA initiative. In each case, the individual character constructs a new persona that is more in line with how they *want* to be seen by the others on the island.

Similarly, as social media researcher danah boyd notes, people "carefully choose what information to put forward" on social network sites.[45] On social network sites, we can use whatever characteristics we want to define ourselves; and while many people tell the truth on these sites, it can be (and often is) a version of the truth that spares some details. By necessity of space, in fact, a social network profile can never reveal *everything* about a person, no more than someone could reveal every aspect of his or her personality in a conversation. But whereas it's difficult to disguise certain physical or personality characteristics in person, online it's as easy as a mouse click to drop 25 pounds or have a sharper wit. When we write ourselves into being online, we show ourselves as we want to be seen. If we include our flaws, it's typically a deliberate inclusion, and if we revise our characteristics, it's usually on purpose. Like Kate, Sawyer, and the rest of the *Lost* castaways, we *choose* who we are online, and how we want to be seen.

Social Connections

Yet, there are aspects of our online self over which we have little control—the conversations that take place within our network of friends and friend-connections. These varied and personal interrelationships provide the most tangible and salient feature of these social networks, as they "enable users to articulate and make visible their social networks."[46] As Donath and boyd have shown, social networks provide "public displays of connection," or friend networks that bolster our online profile.[47] The friends one has on Facebook may reveal more than their profile as to their likes and dislikes, their geographical location, and their life history. We speak more candidly in conversation than our profile creation articulates. When we create our online identities, they are also woven into the networks to which we adhere.[48] We rarely go on Facebook deliberately to meet new people, but rather to network with people we already know—the difference between a "social network" and "social networking" sites.[49] By examining the network of people whom we have "friended," we publically proclaim our social proximity within what boyd calls "networked publics," or spaces where "the imagined collective ... emerges as a result of the intersection of people, technology, and practice."[50] Our online identities are bolstered by a complex interrelationship with others, what Parks shows is the community at the heart of social networking.[51]

The interrelated nature of the characters on *Lost* has been much analyzed.[52] Yet, it's important to restate that it's only through analyzing the *network* of character relations that the narrative is revealed: "following the *Lost* storyline by watching only one character's key episodes in order would provide a viewer with the whole story so far for one character (albeit, still a story of fragments), but the larger story of all the survivors would remain fragmented."[53] Thus, in order to understand the full *Lost* story, it's important to see the hypertextual links between the characters; much like, to understand the full articulation of the online persona, it's crucial to see the influence and participation of an individual's entire social network.

Jack Shepherd represents a key node in the *Lost* social network. The ostensible protagonist of the series, Jack has multiple connections to a variety of characters throughout the show, many of which emerged before his time on the island. For example, Jack's father, Christian, also fathered Claire (through an affair), so Jack and Claire are half-siblings. Sawyer also meets Christian in a bar, who talks about Jack to him before the island. Jack, a surgeon in the world off the island, operated on a woman named Sarah (whom he later married), instead of Adam Rutherford, who came into the operating room at the same time as Sarah, and was the father of Sharon and Boone. Jack also met Desmond before the island, as both were running the steps of a stadium together. Jack and Locke met in the Flash-Sideways season, when paralyzed Locke visited Jack the surgeon. If we take all of Jack's connections pre-, post- and intra-island and imagine that he had created a social network profile linking each of them, we can get a better idea of how complex these relationships become. Further, if we then link one of Jack's connections, Desmond, to their own connections—Desmond knew Kate's father, Desmond met Hurley in a fast food restaurant,

Desmond met Eloise, Daniel's mother, Desmond time travels and visits Daniel in the past, Desmond passed Charlie on the street—then even more connections can be made, and a complete network can be mapped (see *Lostpedia* for more connections, and examples of mapped networks).[54] Ultimately, *Lost* is a social network site, fictionalized, highlighted, and narrativized.

Audience Identification

Like friends in our own social networks, the characters on *Lost* connect in interesting and varied ways. To conceptualize *Lost* as a massive online social network, according to Jennifer Gillan, "encourages fan tracking of characters and stories across multiple media platforms."[55] If viewers can see *Lost* as a giant social network, then perhaps they could see themselves participating in it: as Sandvoss articulates, fans can identify with media objects in a narcissistic self-reflective manner, and media objects form extensions of our sense of self, "who we are, would like to be, and think we are."[56] Just as we sit in the center of our own personal community online, what Baym identifies as a "networked individualism," so too do we sit in the middle of the relationships we see onscreen.[57] The difference is that we are silent members of our television social network.

But even knowing that the relationship between ourselves and the television characters is one-sided, we still feel emotionally connected to them. This type of parasocial identification is used in traditional fan studies as a basis for examining the relationship between a character within a media object and the fan of the media object (see Chapter Three). According to this methodology, fans and other viewers can read meanings into texts not because of what's there, but because of what's absent: the viewer themselves. Readers use these texts like a "mirror of consumption," seeing and reading themselves into the narrative, an imaginative self-representation.[58]

Lost's social network, then, mirrors precisely the social networks we have in our own lives. We only know certain things about characters; we only know certain things about our friends online. At a time where, as Sconce argues, "viewers now have the opportunity to 'inhabit' a given television show for hours on end," the *Lost* audience can picture themselves within this vast social network.[59] Viewers can interact with the narrative through the same cultural lens with which they see their online world: sharing knowledge of social proximity, discussing likes or dislikes or personality traits of characters, and investigating character/friend motivations. Daly points out how today's audience member, what she terms the "viewser," "might find him or herself necessarily less concerned with ... linear narrative than with its meta-construction and intertextual linking. Interactions with digital media have made viewsers familiar with a disordered, hybrid, and unhierarchical navigation of information."[60] Digital technology and transgenic media are creating a networked understanding of human relationships, from which *Lost* and other shows with temporal displacement are learning.

Ultimately, by fostering a connection to social networking, contemporary television programming is tapping into a particularly salient cultural marker and market. Social media have overwhelmingly become the top financial draws online.[61] It is in television's interest to try and cut a piece of this online pie. Mirroring the characteristics of social networking isn't just an aesthetic choice; it's a business decision to develop in parallel with transgenic media characteristics as well.

Narrative Networking

One consequence of this increase in a visual representation of social networking in complex television is the use of *temporal* displacement to construct *spatial* complexity within a narrative. In her discussion of networked publics, boyd discusses how social networks don't just create connections between people, but also, through those connections, create spaces as well.[62] There is a correlation between the social network "spaces" described by boyd, the temporal fracturing of narrative, and the fostering of an immersion in the storyworld of today's complex television. Specifically, social networks online create virtual spaces which then reflect on the participants within them. As online agorae, networks function as places to go, locations where people can "hang out."[63] If I am part of an online social group for people who own dogs, I become identified by that space just as that space becomes part of my identity—I am Dog Owner. As Marie-Laure Ryan points out, many early descriptions of hypertext also created a metaphor of spatial immersion.[64] Early models of Internet communication relied on a spatial metaphor for relating the dialogue of online sociality with non-mediated life (see, for example, William Gibson's term "cyberspace").[65] The conception of social network sites like Facebook and MySpace make them seem like *places* to go, virtual spaces for communication, more akin to Second Life than IMDB. We go on Facebook to see who else is there. Our online network is more than just a list of friends; it is a space where communication occurs, where connections are made.

One major difference between virtual spaces and offline spaces, however, is the way each structures and is structured by time. The temporal landscape of an offline space—mainly synchronous communication—is flipped in virtual spaces, where asynchronous communication is more common. Although real-time chat can occur, and cell phones have muddied the delineation between offline and online spaces, the temporal location of any particular act of communication partly shapes that act. In face-to-face communication, the interaction occurs through multi-sensory inputs, and occurs in real time. But in virtual spaces, we have to continually reconstruct the conversation every time we pick it up; we experience it in "chunks" instead of in bulk. Our reconstruction of a virtual conversation occurs through the same processes that help us structure non-linear narratives and temporal discontinuities.

Similarly, even though it depicts fictional stories, one particularly salient characteristic of temporally displaced television is its "emphasis on storyworld continuity and non-linear narrative."[66] A storyworld is a fully realized space that exists in a narrative, an imagined space that is akin to, but subtly different from, the non-mediated spaces we exist in outside of fictional narratives. Storyworlds are not just environments, but are locations where narrative happens, with characteristics that affect the character and the story. For example, J. R. R. Tolkien's "Middle Earth" is a completely realized storyworld, such that the narrative of the book is almost wholly dependent on the structure of the world so fully realized.

The storyworld of a narrative may be a fictional place (the island on *Lost*, for example) or it may be a fictionalized version of a place that already exists (New York City in *How I Met Your Mother*, for example). Importantly though, the storyworld as it is realized in a narrative is subtly different from the world outside of the story: the New York City of *How I Met Your Mother* is not the same New York City as the one we experience when we visit Manhattan, or even when we watch *Sex in the City* (1998). One function of the storyworld, then, is to set itself apart from the non-mediated world. For example, in the *Supernatural* (2005) episode "Hollywood Babylon," brothers Dean and Sam head to Hollywood to investigate the haunting of a movie set. While taking a tour of the studio, they are told that they are entering the sets from the show *Gilmore Girls* (2000)—a series in which *Supernatural* lead Jared Padalecki also starred. His character's uncomfortable look when *Gilmore Girls* is mentioned as a *fictional* program intertextually echoes the fictionality of *Supernatural* (foreshadowing the fan-based narrative devices of the later seasons). The storyworlds of *Supernatural* and of *Gilmore Girls* necessarily have to be different to understand the intertextual reference, but the fictionality of one reveals the fictionality of the other. In turn, the fictionality of *Supernatural*'s storyworld affects the audience's understanding of the show's narrative too, as well as Padalecki's character.

To develop a deeper understanding of the functions of a storyworld requires new conceptions of narrative, not as just temporally structured, nor as entirely spatially constructed, but as a mashup of the two, a space-time network. It matters not only *where* Sam and Dean hear about *Gilmore Girls*, but also *when* they hear about it (after Padalecki left *Gilmore Girls*). David Herman also indicates this connection, as he states that both temporality and spatialization are two important concepts in narrative structure. His concept of the storyworld indicates the viewer's reconstruction of "not just what happened" in the story, but also "the surrounding context or environment embedding existents, their attributes, and the actions and events in which they are more or less centrally involved."[67] In this way, the storyworld encompasses both "the temporally [and] spatiotemporally structured" events. It's not just television that features an increased discussion of spatial characteristics, for this is "a growing trend … where causality is networked or mapped spatially" across all media.[68]

A storyworld necessarily exists with its own rules, but which mirror those in the non-mediated world. Gwenllian-Jones argues that "the more detail that is available about the [story]world, the denser and more comprehensive it becomes. It assumes the familiar complexity of the material world, sharing its three-dimensional clutter and myriad connective possibilities," in other words, its networked structure.[69] This complexity "furnishes the fictional world of cult television with the gravitas of authenticity."[70] Ryan calls this authenticity "spatial immersion," and notes how detailed offline story environments "need only a few mooring points to take hold of the mind."[71] If, as stated by Janet Murray a "linear story, no matter how complex, moves towards a single encompassing version of a complex human event" then it necessarily "has to end somewhere."[72] But both with temporally displaced narratives and with online narratives, stories can have many endings and can have many different avenues to follow.

Breaks in a narrative timeflow can change not only the character's point of view, but the audience's as well. Mark Stephen Meadows contends that "nonlinear choices need to be allowed within the contexts of the linear story," furnishing a three dimensional view of a storyworld across temporalities.[73] The audience of *Lost* must, along with the main characters, connect those past events with those future outcomes, and those future events with those in the past. For example, the storyworld of *Lost* depends inherently on the geography of the island, and the complex network of characters that exists both on and off the island. But it also depends on when events happen, and the order in which they happen. A key reference to the importance of temporality and spatiality in *Lost* is revealed in the episode "Lockdown," when Locke, trapped in the Swan station, sees a hidden map of the island painted on the wall. The narrative importance of this map emerged from its revelation about the space of the island and its mapping of a network of connections between different DHARMA stations. But its creation over a period of years helps develop the characters and relationships as well, and thus flesh out the character-driven narrative.

Another example of how networked sociality can lead to a storyworld understanding of narrative appears in *How I Met Your Mother*. The five main characters of the show—Ted, Barney, Lily, Marshall, and Robin—form a social network, the connections between them functioning as strong social links. In turn, viewers who feel an emotional connection to these characters may see themselves reflected in the network of the characters: "these friends are just like me and my friends."[74] Yet, the networks go one step further in *How I Met Your Mother*, and actively aid in the construction of a storyworld as well. This storyworld is then textually represented as an aspect of the social network, cohering the audience to the characters as it visually represents the space of a fictionalized New York City.

In most episodes of *How I Met Your Mother*, the characters interact in one of just a few locations: McLaren's bar, Ted's apartment, Lily and Marshall's house, or Barney's office at GNB Bank. Yet, the overarching narrative of *How I Met Your Mother* establishes a larger sense of place within the storyworld of the narrative. The series-long narrative arc finds protagonist Ted discussing with his children the story of how he met their mother—and as of the sixth season, he has not yet met her (nor has it yet been revealed to the audience who she is). But clues have been dropped throughout the series that have hinted towards the identity of the mother. One such clue, the yellow umbrella, is geographically linked to the overarching story, and its existence helps foster a storyworld for the show. As Mittell shows, the revelation of a clue over a long period of time is one crucial element of narrative complexity on television.[75] The yellow umbrella is initially mentioned in the first episode of the third season, "Wait for It," when Future Ted tells his children, "Kids, there's more than one story of how I met your mother. You know ... the thing with your mom's yellow umbrella..."[76] The story begins in the twelfth episode of the third season, "No Tomorrow." In "No Tomorrow," Barney and Ted visit a bar on St. Patrick's Day. Into the bar walks a woman holding the yellow umbrella. The next day, Ted returns to the bar and, finding the umbrella on the ground, takes it with him. In the episode "Right Place Right Time," at the end of season 4, Ted brings the umbrella with him as he walks to get a bagel (and runs into Stella, his former girlfriend and the kids' possible "Mother," Stella). And finally, in "Girls Versus Suits" (Season 5, episode 12), Ted accidentally leaves the umbrella at a girlfriend's house—the roommate of whom, Future Ted confirms, will become the mother.

Each appearance of the umbrella is slight, but the attuned viewer keeps them all in mind in order to construct the larger narrative of how Ted meets his future wife. Each one is related not just to a particular moment in time, but also to a particular place.[77] The umbrella appears four times, and each one is associated with a different New York City location: Future Ted's house, the St. Patrick's Day party, Past Ted's apartment, and the (future) Mother's apartment. It's important not only to understand the connection between the characters that led to these moments, but also to grasp the spatial linkages between them to construct a fuller version of the storyworld. In fictional works, characters are constrained by their surroundings, but also:

> contribute to our understanding of actual-world configurations by returning the projection back to their real-world prototypes. Characters do this by eliciting, from audiences who are bent on critically understanding them, puzzle-solving procedures that become paradigmatic for illuminating aspects of human reality.[78]

The "puzzle" of the umbrella helps lead viewers to "assemble" the larger narrative through engagement with temporally and spatially disparate elements of a storyworld.

The engagement with the narrative, storyworld, and character comes to a head in *How I Met Your Mother* in the sixth-season episode "Subway Wars." In this episode, the friends tease Robin, a non-native New Yorker, about her lack of knowledge about the city. For example, they say, since she's never seen Woody Allen, despite living there for a number of years, she can't be a "real" New Yorker. They present a challenge: arrive at a restaurant by the fastest means possible. Robin takes a cab, Lily takes the subway, Ted goes by bus, Marshall decides to run, and Barney fakes sick and rides in an ambulance most of the way. To map out their journey, Ted pulls out his laptop, opens it to a map website (on sponsor Microsoft's Bing webpage) and shows his friends (as well as the audience) the storyworld of *How I Met Your Mother*: the island of Manhattan. They dart out of the bar, ready to race.

Periodically during the race, the show cuts to that Bing map of New York City, superimposed with all five friends' faces tracking the routes that they take. This map networks the narrative, as short glimpses of each character on route intersperse with the other character's routes, each switch in perspective aided by a glimpse of the map. To further tie together this network, each character impossibly travels simultaneously with guest star Maury Povich, a humorous temporal paradox never explained. "Subway Wars" illustrates how the network of the character intersects with the network of the city: subway lines intersect bus lines, roads intersect sidewalks. Robin leaves her cab and passes Marshall, running. She then runs into Barney driving in a limo. Leaving the limo, Robin starts to take the sidewalk when Barney then runs into Marshall. Visually representing this intersection, the show also uses a five-split screen to show the progress of each character. Furthermore, like Baym's "networked collectivism," other narrative networks contribute to the storyworld: in flashback each character is shown getting disappointing news to illustrate why they each need to win the race (each flashback introduced via Woody-Allen-esque white titles on a black background with clarinet music).[79]

Ultimately, the social network of the characters on this complex television show indicates a connection to the online social networks presented via Facebook, MySpace and other SNS's. Yet, in a move beyond the mere connectability, the narrative of *How I Met Your Mother* creates a realized space within New York City that envelops the characters as well as the viewers within a storyworld. As a direct correlation between the connectability of online social networks and the spaces of that connection, the temporal displacement of the story, visually represented onscreen, fosters a fractured narrative that immerses the viewer within a temporal/spatial fragmentation. Ultimately, through both the storyworld and the fractured temporality, the audience can become spatially and temporally connected to the characters, fostering deeper and more intense identification.

Networked History

If we start to see immense casts of complex narratives on television not as individual *characters* but rather as networks of character *relationships*, then we can begin to see a further connection to online social networks. And further, if we refine those casts to picture them taking place within specific *storyworlds* of narrative action, then we get an even closer correlation to the space of social networking. But in this last section I want to explore a further analysis of the network in relation to online sociality and complex television narrative: the network of *knowledge* such texts engender in the audience.

I've already briefly touched on the way an audience must recognize and keep in mind minor details from episodes across time (as in the Yellow Umbrella trope from *How I Met Your Mother*; Mittell also shows how shows like *Seinfeld* (1990) and *Arrested Development* (2003) rely on an audience's memory of minute facts).[80] Such database-like narratives rely on an active audience being able to create connections between elements, so the presence of networks (both temporal and character driven) is crucial. Indeed, by linking individual units of knowledge in a narrative structure, programs like *Lost* or *How I Met Your Mother* utilize what I've previously described as "narractivity," or the mechanisms of social networks to harness the interactive potentials of the audience to create a narrative.[81]

In the previous two chapters, I've discussed the contradictory mechanism lying at the heart of the intersection of transgenic media and temporal displacement in complex television. On the one hand, transgenic media stabilize our sense of history, collating and saving all our historical data in one point; similarly, complex narratives use temporal displacement to assert the primacy of historical coherence. On the other hand, transgenic media also allow deceit and inadvertent mistakes to creep into our portrayal of our selves online, indicating a historical revisionism or imprecision; similarly, complex narratives use temporal displacement to show the inherent changeability in narrative, the way small events can create massive changes in history.

One reason we can see both contradictory notions as true at the same time is the result of a networked mentality of knowledge within our culture. As I've shown, transgenic media inherently rely on networkization: for example, YouTube video parodies rely inherently on a network sensibility of intertextual connections. Gurney mentions the mashup as characteristic of this sensibility, where video from a number of texts can be juxtaposed to have deeper meaning.[82] The connections between elements create a larger textual meaning than the individual elements can offer. As mentioned earlier, Baym refers to this type of "sum-is-larger-than-the-parts" networking as "networked collectivism," where groups of people use the Internet and related mobile media to create "shared but distributed group identity."[83]

Further, as I've established so far in this chapter, this networked mentality is endemic within complex television as well. As Spigel writes, television has always "recombined narrative materials," making an "increasingly hybrid" medium.[84] Networking connections between elements on television is a stylistic convention of postmodernism, which "tends to reject historical authenticity, mixes styles and genres of television using the technique of *bricolage,* and even draws attention to its own constructedness…in order to frustrate the temptation to derive meaning from an open text." [85] The connection between *Supernatural* and *Gilmore Girls* is a precise example of this postmodern constructedness, although others exist. For example, the actor Jorge Garcia, who plays Hurley on *Lost,* also guest starred on *How I Met Your Mother*. The intertextual references he brought with him from *Lost* abound in *How I Met Your Mother*. On *Lost,* for instance, Hurley is obsessed with the numbers 4, 8, 15, 16, 23, 42. On *How I Met Your Mother,* Steve (Jorge Garcia) appears in the episode "Blitzgiving" and is asked to guess a phone number: he says 481-516-2342. Later, he makes a reference on *HIMYM* to being stuck on an island for what seems like an eternity. The network of references focuses the audience's attention on the constructed nature of both shows, hybridizing the narratives.

Intertextual connections are key to the transmediated nature of many contemporary television programs as well. In 2004, Caldwell mentioned the ways in which *Dawson's Creek* utilized the web to link television and web: by mimicking the look of characters' diaries, computer desktops, and instant message chats, the website dawsonscreek.com effectively kept "viewer-users engaged long after a series episode has aired."[86] *How I Met Your Mother* has continued this trend by intertextually connecting its series to a blog, purportedly written by womanizing Barney. Each entry on his blog is usually in response to what happens during each of the episodes. Besides being a usually humorous take on the events of the episode, the blog also becomes part of the show's text: Barney mentions it in the pilot episode and again in the episodes "Okay Awesome" and "Rabbit or Duck." Such links between blog and show are an example of transmediation described by Jenkins, where different media present different aspects of a single narrative.[87] Transmediation functions precisely through the same mechanism as networking. By linking narrative across media boundaries, the transmediated text becomes understandable because of the connections between elements. Transgenic media function similarly to connect narrative, history, and knowledge, by linking knowledge across textual boundaries. Transgenic media elucidate multiple sides of any issues. If we previously saw both the concept of "history" realized through complex television as entirely writable and stable *as well as* the concept of "history" as entirely infinite and unstable, then the network offers us a way to view these two contradictory ideas together. By seeing the networking as a *force,* not just a series of connections, we can see the active power of the creation of relationships. And that connective power coheres disparate thoughts. Quite simply, because we *can* see connections between disparate beliefs, ideas, or arguments, we *create* a networked understanding of everyday life.

By viewing our personal relationship with social networking, with wikis, with online videos, or with other transgenic media, we see the multiple connections, the "networked collectivity" or our lives.[88] And the network metaphor gives us another way of understanding these connections: as parallel tracks in our knowledge creation. We can recognize the constructed nature of history and also its ultimate stability online at the same time through the same process that allows us to see New York City as a real place and as a fictional place simultaneously, or to see the *Lost* characters as fictional but meaningful. Television doesn't just offer us a glimpse of another world; it also offers us a metaphor to understand how to live our own lives. As Mittell shows, "If we want to truly understand how television is consumed and understood in today's world, we need to pluralize the medium, considering how it is used on a range of screens, through an array of interfaces, and placed within a diverse set of cultural rubrics."[89] Yet, to understand our consumption of everyday life, with all its complexities, awkwardness, and contradictions, we also have to understand the way we view the networking metaphor of complex television.

Conclusion

In the shift from a Web 1.0 world to a Web 2.0 world (or a Web Commons world, as I have previously called it[90]), we've seen increased attention in the way individuals are perceived to have power and activity in our culture. Although much has been made of the individualization of the web (the "i"-ification of so many Apple products, or focusing on "you" as *Time*'s 2006 person of the year), I believe the strength of transgenic media lies in its concentration on networks as a structuring element.[91] We live at a time when the power of the network is only beginning to be understood. As I was writing this, people in the Middle East and North Africa mobilized through transgenic media to seek political change. Thousands sought regime change in Egypt via Twitter and Facebook. Money was raised for multiple charities, political campaigns, and relief causes via mobile messaging, Facebook posts, and Twitter. Human social networks, enhanced by technology, help create social change.

The mediated experience of our lives has changed. I don't mean this in a purely deterministic manner, but I also don't think it a stretch to see that with the massive influx of new technology comes societal changes. For example, the rise of the "age of networks" has helped foster a change in the social construction of time.[92] As we've seen, we experience *time* differently because of our interconnected, networked lives. These changes are also represented in the way television narratives have shifted in the first decade of the 21st century. By focusing on the "network" as a metaphor and as a symbol for 21st-century life, television not only reflects back at us the experiences of our everyday life, but also helps shape those experiences so that we can understand them and place them in a knowable context.

It's not just television that's making this intuitive leap towards extreme networkization. In the print medium, scores of novels have emerged which feature networked plots, complex character inter-relationships, and social networking characteristics. For example, Dan Chaon's *Await Your Reply* (2009) tells three seemingly unrelated stories about the construction of identity and its effect on individuals and groups. Chaon's novel appears to draw on Goffman's identity performance characteristics as much as it draws on the characteristics of profile creation when the main protagonist constructs new identities for himself. Further, Kevin Brockmeier's *A Brief History of the Dead* (2007) sees all of human life as a social networking profile and posits that you don't pass out of purgatory until everyone on Earth that once knew you has forgotten you or dies, effectively using the social network to connect individuals from life to death. With characters linked with both strong and weak ties to others, the parallel to social networking is made clear when the last human on Earth, struggling to stay alive, remembers previous acquaintances, including a former best friend, her parents, and a blind man she once passed on the street. Kept "alive" by the network of connections, each character is represented through their relationship with others.

Even in the video game medium we see similar temporal situations. 2008's *Braid* finds the main character playing with time—reversing, stopping, and condensing—as a major aspect of the gameplay. Players complete levels, but must return to replay again and again to solve the mystery of the game. In Silicon Knight's 2002 *Eternal Darkness: Sanity's Requiem*, the player must play through the game three times, making minute variations on each journey through the game to effect monumental changes in the outcome. Only then, after networking the three narratives together, does the player reach the true end of the game, where the massive Lovecraftian God, released from captivity, describes the networked "truth" of the world: "All at once, separate and simultaneous...the universe is made of many time streams, many possibilities, all in harmonious synchronicity."[93]

Ultimately, the power of the network as "harmonious synchronicity," as a series of relationships, as a space, and as a metaphor for understanding how we as a species relate to knowledge and history becomes apparent in the way that our everyday media reflect network characteristics. The use of temporal displacement in contemporary complex television is the most telling representation. By playing in time with memory, character, and narrative, today's television portrays our modern-day temporal disorientation. Yet, the complexity of television is belied by its popularity, and by offering a narrative ending to these temporal paradoxes, these television texts help foster a deeper and more meaningful understanding of our culture. It's not that changes in television have led to a greater acceptance of temporal discontinuity in our own lives, but rather that in its reflection of our quotidian lives, television captures those fleeting moments, and then gives them back to us.

Notes

1 Yu, 226.

2 Of course, this is a central facet not only because of the significance of the social but also because the nature of conflict depends on this interaction.

3 Johnson; see also Todd Gitlin, "'Make It Look Messy: Striking a Note with *Hill Street Blues*," *American Film* (Sept 1981): 45–49, 76.

4 Allen, *Speaking*, 86, my emphasis.

5 Allen, "Audience-Oriented," 108; see Mittell, "Narrative Complexity."

6 Mittell, "Narrative Complexity," 32.

7 Johnson, 116.

8 Ibid., 108.

9 Holly Haynes, "As *Lost* Ends, Creators Explain How They Did It, What's Going on," *Wired* 18, no. 5 (19 April 2010), http://www.wired.com/magazine/2010/04/ff_lost/5/ (accessed 14 Mar 2011); "Lost Connections," *Lost* Season 2 DVD (Los Angeles: Buena Vista Home Entertainment, 2006).

10 See George Landow, *Hypertext 3.0: Critical Theory and New Media in an Era of Globalization* (Baltimore, MD: The Johns Hopkins University Press, 2006).

11 Yochai Benkler, *The Wealth of Networks* (New Haven, CT: Yale University Press, 2006); Jan van Dijk, *Network Society: Social Aspects of the New Media,* 2nd ed. (London: Sage Publications, 2005); Zizi Papacharissi, ed. *A Networked Self: Identity, Community, and Culture on Social Network Sites* (New York: Routledge, 2011); Castells, *Internet*; and Manuel Castells, *The Rise of the Network Society: The Information Age: Economy, Society, and Culture Volume 1,* 2nd ed. (Malden, MA: Wiley-Blackwell, 2009).

12 van Dijk, *Network,* 2.

13 Castells, *Rise of the Network Society*; see also Daniel Chamberlain, "Media Interfaces, Networked Media Space, and the Mass Customization of Everyday Space," in *Flow TV: Television in the Age of Media Convergence*, ed. Michael Kackman, Marnie Binfield, Matthew Thomas Payne, Allison Perlman, Bryan Sebok (New York: Routledge, 2010), 17; also Castells, *Internet.*

14 John Fiske and John Hartley, *Reading Television*, 2nd ed. (New York: Routledge, 2003), 85.

15 Margaret Somers, "The Narrative Constitution of Identity: A Relationship and Network Approach," *Theory and Society* 23 (1994).

16 Williams and Keady, 163.
17 Somers, 606.
18 Ibid., 613.
19 Ibid., 614.
20 Ibid., 616.
21 Shirky, 1–14.
22 Bruce Simmons, "'Supernatural' Episode 'The French Mistake' Review and Misha's Tweet," *Brusimm* (26 Feb 2011), http://brusimm.com/2011/02/26/supernatural-episode-the-french-mistake-review-mishas-tweet/ (accessed 14 Aug 2011).
23 Daly, 86.
24 Benkler's analysis mirrors a similar conception of the emergence of the symbolic in Jean Baudrillard, *For a Critique of the Political Economy of the Sign* (1968), trans. Charles Levin (US: Telos Press, 1981).
25 Benkler, 3.
26 Ibid., 3.
27 Ibid., 11.
28 Tyron, 111.
29 See Booth, *Digital Fandom,* 48, for a discussion of blog comments.
30 Johnson, 70.
31 Daly, 82.
32 Ross, 173.
33 Ibid., 175.
34 Ibid., 204–06.
35 Johnson, 115, although he is speaking specifically about the show *24.*
36 "MySpace," *Wikipedia* (2011), http://en.wikipedia.org/wiki/Myspace (accessed 16 March 2011).
37 Andrea L. Press and Bruce A. Williams, *The New Media Environment: An Introduction* (Malden, MA: Wiley-Blackwell, 2010), 64.
38 "Press Room," *Facebook* (2011), http://www.facebook.com/press/info.php?statistics (accessed 16 Mar 2011).
39 Bruns, 319–20.
40 Zizi Papacharissi, "Conclusion: A Networked Self," in *A Networked Self: Identity, Community, and Culture on Social Network Sites,* ed. Zizi Papacharissi (New York: Routledge, 2011), 307.
41 Goffman, 252.
42 Ibid.
43 Papacharissi, "Conclusion," 304; also Booth, *Digital Fandom.*
44 Ibid., 307.
45 danah boyd, "Why Youth (Heart) Social Network Sites: The Role of Networked Publics in Teenage Social Life," in *Youth, Identity, and Digital Media Volume,* ed. David Buckingham (Cambridge, MA: The MIT Press, 2007), http://www.danah.org/papers/WhyYouthHeart.pdf (accessed 10 Jan 2008), 12.
46 danah boyd and Nicole Ellison, "Social Network Sites: Definition, History, and Scholarship," *Journal of Computer-Mediated Communication* 13, no. 1 (2007), http://jcmc.indiana.edu/vol13/issue1/boyd.ellison.html (accessed 01 Jan 2010): ¶6.

47 Judith Donath and danah boyd, "Public Displays of Connection," *BT Technology Journal* 22, no. 4 (2004).
48 boyd, "Why Youth," 12.
49 boyd and Ellison.
50 danah boyd, "Social Network Sites as Networked Publics: Affordances, Dynamics, and Implications," in *A Networked Self: Identity, Community, and Culture on Social Network Sites*, ed. Zizi Papacharissi (New York: Routledge, 2011), 39.
51 Malcolm R. Parks, "Social Network Sites as Virtual Communities," in *A Networked Self: Identity, Community, and Culture on Social Network Sites*, ed. Zizi Papacharissi (New York: Routledge, 2011).
52 Gillan, 156–59.
53 Ibid., 159.
54 "Character Connection," *Lostpedia* (2011), http://lostpedia.wikia.com/wiki/Character_connections (accessed 04 Mar 2011).
55 Gillan, 2.
56 Sandvoss, *Fans*, 96.
57 Nancy Baym, *Personal Connections in the Digital Age* (Cambridge, UK: Polity Press, 2010), 90.
58 Sandvoss, *Fans*.
59 Sconce, "What If?" 110.
60 Daly, 92.
61 Susanne Craig and Andrew Ross Sorkin, "Goldman Offering Clients a Chance to Invest in Facebook," 02 Jan 2011, http://dealbook.nytimes.com/2011/01/02/goldman-invests-in-facebook-at-50-billion-valuation/ (accessed 02 Jan 2011).
62 boyd, "Social Network Sites as Networked Publics."
63 See, especially, boyd, "Why Youth."
64 Ryan, 260.
65 See, for a discussion of "cyberspace" as a term, David J. Gunkel and Ann Hetzel Gunkel, "Terra Nova 2.0—The New Worlds of MMORPGs," *Critical Studies in Media Communication* 26, no. 2 (2009).
66 Mittell, "Previously On," ¶11.
67 David Herman, *Story Logic: Problems and Possibilities of Narrative* (Lincoln: Univeristy of Nebraska Press, 2002), 14.
68 Daly, 96.
69 Gwenllian-Jones, "Virtual Reality," 93.
70 Booth, *Digital Fandom*, 97.
71 Ryan, 127.
72 Murray, 136.
73 Mark Stephen Meadows, *Pause and Effect* (Berkeley, CA: New Riders, 2003).
74 As I pointed out in *Digital Fandom*, 153–78, some viewers do take the actualization step, and create social network profiles for these characters, with which to roleplay or interact with other fans.
75 Mittell, "Narrative Complexity."
76 Carter Bays and Craig Thomas, writer, "Wait for It," *How I Met Your Mother* (Los Angeles, CA: CBS, 2007).

77 See Johnson, 88–89, for a discussion about these long-term jokes.

78 Newman, 73.

79 See Baym, *Personal Connections,* 91.

80 Mittell, "Narrative Complexity," 35.

81 Booth, *Digital Fandom,* 103–26.

82 David Gurney, "'It's Just Like a Mini-Mall!' Textuality and Participatory Culture on YouTube," in *Flow TV: Television in the Age of Media Convergence,* ed. Michael Kackman, Marnie Binfield, Matthew Thomas Payne, Allison Perlman, and Bryan Sebok (New York: Routledge, 2010), 41–43.

83 Baym, *Personal Connections,* 91.

84 Spigel, 4.

85 Adrian Page, "Postmodern Drama," in *The Television Genre Book,* 2nd ed, ed. Glen Cleeber, Toby Miller, and John Tulloch (Houndmills, UK: Palgrave Macmillan, 2008), 54.

86 Caldwell, "Convergence Television," 51.

87 Barney's blog can be found at: http://www.cbs.com/primetime/how_i_met_your_mother/community/barney_blog/index.php; the *How I Met Your Mother* Wiki can be found at: http://how-i-met-your-mother.wikia.com/wiki/How_I_Met_Your_Mother_Wiki; Jenkins, "Transmediation 101."

88 Baym, *Personal Connections,* 91.

89 Jason Mittell, "TiVoing Childhood: Time-Shifting a Generation's Concept of Television," in *Flow TV: Television in the Age of Media Convergence,* ed. Michael Kackman, Marnie Binfield, Matthew Thomas Payne, Allison Perlman, and Bryan Sebok (New York: Routledge, 2010), 53.

90 Booth, *Digital Fandom.*

91 Lev Grossman, "*Time*'s Person of the Year: You," *Time* (2006), http://www.time.com/time/magazine/article/0,9171,1569514,00.html (accessed 18 Sept 2008).

92 See van Dijk, *Network,* 157–61.

93 *Eternal Darkness: Sanity's Requiem* (St. Catharines, CA: Silicon Knights, 2002).

Section IV: Conclusion

The chapter in this section concludes the book by examining the twinned futures of television and transgenic media. Both television and transgenic media employ temporal displacement, and as I've shown in this book so far, also teach the audience the skills needed to situate themselves within a complex, postmodern culture.

The realization of temporal displacement can be debilitating, as it registers as a break with the notion of historical identity and place. With this break, though, comes a liberatory aspect—a freedom to explore, join, create, and exist outside of societal and cultural absolutes. By demonstrating mastery over the temporality of a narrative television show, active viewers can satisfyingly return to their digital technology, with all its insubstantial temporality in check. In this final chapter, then, I want to discuss how the interaction between television and transgenic media helps express a moment past "postmodernism," and look at how today's media environment may, in fact, be speaking outside traditional theory.

Seven

The Future of (Mashup) Television?

> *How do we know what it means to perceive an event as presently occurring, rather than as a memory of a past event? How can we tell present from past? And how do we move the infinitesimal window of the present through the viewfinder at such a constant rate? Why can we see a faraway snow-tipped mountain range, or a jet taking off, or the moon, or the sun, or stars, and not an event that took place a moment ago, let alone a month ago, a year, thirty-three years ago?...Maybe it's because we need to be able to do so, for our survival...our physical ability to understand time has been honed by evolutionary pressures to select for traits useful for survival, in all aspects, and time perception is no exception or special case or even magical or mysterious case.*
>
> —Yu, How to Live Safely in a Science Fictional Universe, 178

We live eternally in temporal distortion. It is because of our need to linearize our life that we elide the temporal displacement that is all around us. As Yu points out, if we can see the stars, why not the past? But those stars we view above our heads *are* the past. The light from the stars shows us what they *were,* not what they *are.* They twinkle with images not from the present, but rather offer glimpses of a time long ago, their light moving towards us as fast as possible, but slower than time. We are so far from the sun that it takes its light approximately eight minutes to reach us. The sun could be destroyed and we would have the time between commercial breaks before we saw it. Or, in a more intimate sense, temporal displacement occurs in the split millisecond between the intonation of a voice and the aural reception of that message: we can never have an authentically *synchronous* dialogue. It may be imperceptible, but temporal displacement occurs every second of every day in every conversation.

So it should come as no surprise that as our technology has become more personal, more intimate, and more ubiquitous, it also starts to mirror some of the temporal effects that we experience in our everyday lives. Take, for example, a mobile phone conversation, which may sound synchronous until we watch the mouth of our partner as we hear her voice on the phone. The short lapse in voice transmission means that every mobile phone conversation occurs slightly discontinuously with the speaker.

As Yu describes, we create a sense of linear time in order to survive. As I've described in this book, our transgenic media—the online media that includes new audience interaction mashed up with traditional communicative characteristics—refocus attention on the non-linear characteristics of time within a cultural system. For contemporary postmodern viewers, the world is itself seen as a constant displacement from temporality: we have entered what is, as we have seen, "a series of pure and unrelated presents in time."[1] One effect of transgenic technology is to separate out the time it takes to communicate from the process of communication. The sheer instantaneousness of communication and the inherent multi-linearity of transgenic media create a new sense of temporality for audiences. Of course, I'm not saying that transgenic media have confused people's sense of time, or that audiences understand complex temporality *because* of the Internet—that form of determinism is far too simplistic to address the complex interaction between these diverse technologies and social processes. However, the interaction between contemporary culture's use of temporal displacement and the rise of transgenic technology indicates an emphasis on a larger issue in society: namely, a shifting notion of the cultural response to time and temporality in general.

As we've seen in this book, one of the most salient media technologies that reflects this displacement is television. Television is still one of the most popular media technologies: although we like to think the Internet has taken over all entertainment (and for some people, it has), the television still remains the most watched medium in our culture.[2] It also functions as a form of *personal* entertainment, one with which the audience can readily identify. As Lisa Parks notes, "if we're really concerned about what the 'next generation' of television might look like, we ought to see and imagine ourselves as part of it."[3] It's the future of television that concerns me here; however, prognostications of the future of any media technology are fraught with difficulty. By the time this book arrives on the bookshelves, any predictions may already be laughably redundant (how's that for temporal displacement?). That being said, I believe the future of television as a medium is secure. Televisions will get bigger, and smaller, but most people will continue to have moderately-sized machines (although what "moderate" means culturally will shift and change, becoming both larger and thinner). Just as Gripsrud sees "digitation," or the expansion of digital technology, as "both one of the causes of what seems to be a major restructuring of social life…and one of the effects of this ongoing process," I see digital technology affecting the television.[4] But what is to come? What is the future of television in this digital environment?

I want to make the argument that the key to looking at the future of television is not through changes in technology or industry, but through shifts in the aesthetics and structure of television content. The hegemonic impulses that control and delimit the types of television that is produced in our culture are massive and insidious: they help determine, just as they are determined by, the pace of cultural change engendered by more interactive and user-generated media online. This isn't to say that all media online are free of ideology: far from it. Indeed, the democratic impulses of the online world are tempered by political, social, and industrial changes. The argument here isn't that one technology is more freeing than another, but rather that the intersection of the two technologies demonstrates profound changes happening in our society at multiple levels: changes that have not yet all come to fruition.

Technology changes too quickly to keep pace with book publishing: perhaps the television of the future will aggregate different media technologies within one device. Such an idea may already be here thanks to the iPad. Already, digital apps dominate some television screens, allowing viewers to remix their own "flow" from content taken from the web, from Netflix, or from Hulu. Perhaps 3D TV will dominate, or maybe we are but a few years from Holographic technology. In terms of the industry, we may see a shift in ownership (more and more consolidation into a very few industrial hands) or the purchase of television studios by ever more powerful corporations. This consolidation will probably mirror the rise of individualized, personalized programming online that is independent and caters to the hundreds of thousands. But these are industrial, technological changes that others have more ably hypothesized before me.[5]

The aesthetic changes, however, will be most meaningful, most relevant and most immediate to everyday life.

The Future of Television: Media 2.0

One way of looking at the aesthetic future of television is to see its transformation over the past decade. Using a "Media Studies 2.0" methodology, as I have done in this book, allows us to see individual texts as components of larger cultural shifts in media. In many ways, Media Studies 2.0 allows scholars and practitioners of media to deal with the issues of a

> fluid, connected, always-on, digital ecology of hybrid intercommunication forms, messages, content and activities—personalized and individually and immediately available; controllable and manipulable at will and feeding-into, promoting or giving rise to personal production, content, and meaning-creation.[6]

Gauntlett argues that the media studies landscape "has changed significantly in the past fifteen years, so that the traditional textbook categories of 'film,' 'television,' 'newspapers,' etc., also with 'the Internet' in a parallel but separate category, are increasingly redundant."[7] Media Studies 2.0 shows not only the convergence of technology, but also the increased convergence of production and consumption, forcing a shift in the varied ways we interpret our media. For example, how do we classify online remix artists, who don't "create" music, per se, but who remix content as an act of creation in and of itself?

This Media Studies 2.0 mentality, however, is not new. Seeing the media environment not as a string of individual technologies (the "siloed" study of media) but rather as a comprehensive *mashup* of them all, is the basic definition of televisual studies.[8] Indeed, as different technologies are invented or become more popular, we find new ways of talking about older technology. The uniqueness of individual media technologies is a fallacy Thorburn and Jenkins call "media purity."[9] They argue that any medium analysis must take into account other media as well, as "medium-specific perspectives may limit our understanding of the ways in which media interact, shift and collude with one another."[10] We cannot know one technology without an understanding of other media against which to compare it: after all, as Derrida has reminded us, we didn't even *know* the concept of speech until we had writing.[11]

Television is the primary example of this fluid sense of media analysis. In previous scholarly literature, television has been compared to every form of media from oral traditions to the videographic. For example, Fiske and Hartley describe its "bardic" function, harkening back to the oral tradition which is also implied in the "secondary orality" articulated by Ong.[12] They argue that tools of literary discourse don't fit television: "Every medium has its own unique set of characteristics, but the codes which structure the 'language' of television are much more like those of speech than of writing."[13] Indeed, Thorburn argues that television is, at its most basic, a storytelling device, and for this reason it functions "as an instrument for continuity as well as change, a communication system devoted most of the time to entertaining as many of us as possible with stories and fables that earlier media and story systems had told before."[14] Using the "oral" characteristics of television as a starting point for media analysis allows scholars like Welch to link television with classical storytelling traditions.[15]

In contrast to Fiske and Hartley, others have compared television specifically to literature. For example, Horace Newcomb finds that television, "like the literary form, can offer a far greater sense of density" of material.[16] Also comparing television to literature, Glen Creeber notes that television scholarship can "explain the multilayered intertextual possibilities" of serialized television as a theory of the novel.[17] Expanding on television's connection to the written word, Marshall McLuhan famously compares television and comics as "cool media," those that require audience participation in order to determine meaning, a discussion mirrored by Scott McCloud's analysis of graphic novels and comic books.[18]

Even recently, television has been compared to other media. In *Televisuality*, for example, John Caldwell sees television as the synthesis of two great trends in media technology, the cinemagraphic and the videographic.[19] On the one hand, shows like *Twin Peaks* (1990), he argues, demonstrate a commitment to cinematic values and aesthetics. On the other hand, a videographic iconography displays the fast and spirited nature of video technology in television graphics. Today, we might augment his analysis with the inclusion of digital graphics and link the television to the rise of the home computer.

And in this book, I too have connected television with another medium, this time the online, social, interactive, and participatory media that I call "transgenic media." It is transgenic because it takes something familiar—a traditional communication style and now-familiar digital technology—and inserts a new component—audience participation. That is, transgenic media is not simply a "Web 1.0" technology, but rather integrates a greater sense of audience participation into the development of new technology. When television takes on the aesthetics of transgenic media, I term this Mashup Television, after the symbiotic nature of relationship between the two media.

A "Media Studies 2.0" analysis might find that *because* television mediates everyday life, it becomes the medium most integral *to* this mediated life. Television reflects the mediation of real-life back at us, in what I've previously called a demediation of content.[20] As Creeber has suggested, today's long-form serialized television has a "tendency towards narrative complexity" and with its

> ability to produce a 'flexi-narrative' structure that mixes a number of narrative levels together (frequently without offering any neat resolutions), [it] is arguably better able to reflect and respond to the increasing uncertainties and social ambiguities of the contemporary world.[21]

In his view, television is a postmodern medium, one that effectively portrays the fragmentation of meaning throughout contemporary thought. I've also used this postmodern analysis in this book to offer one particular mode of discussion for temporal displacement, but I want to make a turn here to see what other forms of thought might elucidate contemporary television theory. Specifically, I want to turn from a postmodernist model of television to a "remix" theory of television content and show how television remixes *taste*.

The Future of Television: Remix Theory and Taste

In his essay for *Remix Theory*, Lev Manovich calls our current epoch a "remix culture," following on from the "postmodern" culture of the 1980s.[22] A remix combines samples from older media to create a mashup of content and aesthetics. Unlike the postmodern notion that temporal displacement on television reflects a schizophrenic sense of personal identity, a remix culture might argue that temporal displacement instead stems from the aesthetics of sampling, appropriation, quoting or remixing of content across temporal boundaries.

Television fits solidly into remix culture by taking our everyday life and "remixing" it to mediate it back to us. Remixing is an active process, always engaged with mediation. The remix thus exemplifies a cultural shift discussed by Williams, away from analyses of individual products and towards examinations of larger cultural practices.[23] Television itself is a constant remix—a medium always in flux. That is, much of what we see on television is a remediation of content we've seen elsewhere (clip shows, news reports, repeats). Every audience member also necessarily brings his/her own background into understanding media objects, remixing the content onscreen with his/her own history and background. Television also remixes the aesthetics of other media to ease comprehension; for example, using cinematic, videographic, or literate aesthetics as shorthand for genre or symbolism.

Television also "remixes" two different styles: it exists in an existential duality as both the Minnowian "vast wasteland" and what Johnson describes as an intellectually stimulating medium.[24] Flip through the channels and encounter simplistic, easily-digestible entertainment as well as complex, thoughtful productions. The ability to mashup two mentalities, two separate "tastes," is what has made television the dominant medium in our lives. Taken to its logical end, therefore, this duality leads us to see that contemporary television also complicates this notion of taste. By looking at television through a generative "remix" lens instead of pessimistic "postmodern" one, I hope to demonstrate an understanding of how taste filters through mashup television.

Traditionally, the notion of taste rests on a linear sense of temporality, not the multi-faceted system demonstrated through temporal displacement. But in a remix culture, a redefinition of that marker of value indicates the influence of transgenic media within a mediated environment. In qualitative matters of taste as they relate to judgments about the quality of cultural objects, Pierre Bourdieu's *Distinction* remains the classic work in the field.[25] Bourdieu assigns all aspects of a text's quality to the social class of the audience, not to the object itself: "the capacity to confer aesthetic status on objects that are banal or even 'common' (because the 'common' people make them their own, especially for aesthetic purposes)...[reverses] the popular disposition which annexes aesthetics to ethics."[26] A work that previously may have been thought value-less in a culture ruled by aristocrats may become re-valued when discussed by/in a larger population.

Bourdieu's theory represents a turnabout from previous conceptions of how value was "earned" in society, many of which depended on a temporal sense of history. The older something was, the more valued it became. As Bourdieu describes, "Legitimate manners owe their value to the fact that they manifest the rarest conditions of acquisition, that is, a social power over time which is tacitly recognized as the supreme excellence: to possess things from the past…is to master time."[27] In contrast, to describe the way that value is ascribed in a social system, Bourdieu argues that the mode of appreciation depends not so much on the idea of a "value-through-time" model of taste (which high culture enthusiasts might support), but rather on an immediate, social space created and defined by a structure (habitus) that itself is determined by the population's conditions of existence.

Asserting the quality of a particular aspect of television programming is problematic given the multifaceted view of taste asserted by Bourdieu. What is quality television to one person may be dismissed by another as mere fluff. Perhaps one future of television would see this issue of "quality" expanded, as the continued appreciation of television textuality may synthesize past and present television "tastes." For example, a 2010 commercial for the Volkswagen Jetta remixes digitized clips of Gene Kelly and Donald O'Conner doing a sitting dance in the backseat of a Jetta.[28] Similar remixes have been made previously; for example, Elton John "performed" for Humphrey Bogart, James Cagney, and Louis Armstrong in a commercial in 1991, and Fred Astaire, Lucille Ball and John Wayne have all had their appearance digitally inserted into commercials for Dirt Devil vacuum cleaners, the California Lottery and Coors Light, respectively. As Broydo argues, it's becoming more common to use the images of dead celebrities to hawk products.[29] The remix of past celebrity with present product becomes more than just a mashup of technologies—the analogue technology of the past meets the digital technology of the present—but also becomes a mashup of *tastes*. On the one hand, the fact that many see this as "bad taste" (Marshall reproduces some negative Tweets about these ads), and yet the Volkswagon to many symbolizes "good taste" is yet another reminder of the mashup between the two on television.[30] On the other hand, these commercials seem to mashup past and present, death and life, entertainment and commercialism. Taste has never been more fluid.

New Technology, New Television, New Audience?

Audiences are becoming more attuned to temporal discrepancies; more used to jumps in narrative. Thanks in part to the success of the DVDs for *Lost, Family Guy, 24, Firefly, Arrested Development,* and others, more shows are being made which rely on a particular mindset of the audience to understand, follow, and discuss complex narratives. Additionally, in recent years there has been an explosion in the availability of consumer-priced electronics to make even the most techno-phobic user a home video pioneer. From TiVo to HDTV, from Blu-Ray to Plasma screens, the sheer influx of new technology creates a disarmingly high volume of "technology" to purchase. And as rapidly as technology changes, so too does the type of "thinking"—the intellect—of media consumers change.

Transgenic media technology lies at the forefront of a rapidly changing mindset in consumers. As Johnson points out, new media has encouraged "more intellectually demanding" viewers—audience members whose intellects have been stimulated and advanced by the demands that new media places upon them.[31] Jenkins argues that the television industry relies on this new, digitally savvy audience.[32] Audiences are becoming more familiar with interactive technology like the DVD, and media producers of these new technologies declare that the "text" is becoming easier for the audience to grasp (although, as Brookey and I have pointed out, so-called "interactivity" is often just another mechanism used to sell merchandise).[33]

The transgenic media I've discussed in this book can also be seen as interactive, although this interactivity is qualitatively different from the type experienced on the DVD. For example, on a wiki, anyone can contribute to a webpage: authoring is limited only to those with an Internet connection and an idea. This ability to alter or change anything on a wiki is part of what makes a wiki appealing: a web page can be almost contemporaneous with current events, as users can post real-time reports about situations around the world. We can see this most fully in wiki pages that spring up about natural disasters: the Wikipedia entry for the 2010 Haiti earthquake developed literally overnight; and for many weeks afterwards was an authoritative aggregator of collated news items about the event and the rescue attempts. Yet, this interactivity has a counterpoint: at the same time, the fact that anything can be changed on a wiki means that no one contribution has any sense of permanence. Unless a contributor is willing to monitor whether his/her writings have changed, the possibility always exists that one's contributions can be deleted. This sense of impermanence contrasts strongly with a historical notion of the permanence of writing, as Bolter has suggested.[34]

Similarly, the idea of social network sites, on which users can friend other people and contribute to a network of linked users, also relies on the notion of impermanence. The home page for any user is a mashup of their own personal information (which they can update or change at any time) as well as the input of their friends. As I have previously pointed out, any social network site is constructed from the input of others, as friends leave notes, memos, videos, thoughts and comments on each other's pages.[35] No page is permanent, as the output of friends will change the home page continuously. In an era when our digital identity can be based on what we put on a social network site, the fact that the site can be in continuous flux quite literally means that our digital personality is never constant.

Further, although a blog might be considered a simple online diary on which users can post their thoughts, the technology of the blog is a great deal more complicated.[36] The blog actually consists of two different textual areas: the post, which is usually authored by one person, and the comments, which are usually authored by many. Each comment builds upon the blog text: the complete blog grows with every comment. However, amid the many comments, the voices of many may drown a single voice out, or the author of the blog post may decide to delete it entirely.

The possibility that one's unique contribution to a digital text may be ignored or removed highlights the insubstantiality of digital technology. Users may exalt the ability of digital technology to be updatable and uniquely interactive, but the possibility of the elimination of any one contribution may also be terrifying, and brings to a head the notion of a remix-like diaphanousness. By exemplifying these interactive texts, through the blog-like *Life on Mars* and *Ashes to Ashes*, wiki-like *Doctor Who*, and social network-like *Lost* and *How I Met Your Mother*, television not only reflects the uncertainty of the contemporary media landscape, but also offers viewers a method of appreciation and acceptance of this cultural ambiguity.

Television is not the only medium that has adjusted its textuality to take into account changes in the media environment. Cinema texts like *Pulp Fiction* (1994), *The Usual Suspects* (1995), *Run Lola Run* (1998), *Memento* (2000), and *Donnie Darko* (2001) also play with narrative time in unusual ways. Video game narratives like *Eternal Darkness* (2002) demonstrate temporal complexity.[37] New technology like DVDs and MMORPGs create interactive temporalities. Film scholar Costas Constandinides builds on Gauntlett's argument to state that it is now impossible to separate media today, as all are constructed using digital technology in a ubiquitous media environment. In his study of "post-celluloid" film, he uses the metaphor of "adaptation" as a vehicle for exploring changes that new media have engendered in the way we view older media.[38] Every text becomes "one artifact in a multimedia, interactive cinematic experience involving websites, video games, DVDs, contests, mobile media, and so on."[39] Marsha Kinder argues that traditional media mirror the aesthetics of digital technology, including its database structure.[40] Further, Derek Kompare shows how the filmic, televisual, and digital industries connect through the release of television series on DVD.[41] He argues that through DVD technology, television is changing its "business models, regulatory structures, programming forms, and modes of viewing."[42] Kompare shows how the DVD changes our conception of the *content* of television to the *object* of television, leading to a vast commercial enterprise that objectifies television programming for purchase.[43] These new media technologies, and other, more mobile technologies, lead to what Brooker has shown is "our increasing engagement with the world as data, through digital technology, [which] involves new vocabulary, gestures, conventions and conceptual models: new ways of seeing, acting and thinking."[44]

These new ways of seeing, acting, and thinking reflect on the modes of viewership enacted by today's audiences. Attuned to digital technology, audiences today watch their older media with an analytic eye to the new. Although such changes can be seen in film, DVD, and video games, television is still the most visible, most watched, most familiar media technology in our lives today, and the impact of these temporally displaced complex narratives is far-reaching.

Conclusion: (New) Mediated Communication

Perhaps, then, what has been most significant about this shift in television narrative is how it heralds new modes of mediated communication. In this book I've focused on two different areas of emphasis within televised narratives, character and narrative. I've also articulated a shift in the way that these different areas interact with the already-complex audience of television, one that imagines the "mashup" as a metaphor for larger changes in our culture.

Tying together all these changes, I've articulated how television programming utilizes a structure of temporal displacement as a unifying theme across multiple shows. Temporal displacement, or the complexity engendered by narrative shifts in time including flashforwards, flashbacks, time travel, and changes in the protagonist's memory, heighten the spectacle of a television narrative. By then amplifying and mirroring an audience's acceptance of a cultural temporal discontinuity, television illustrates the emergent complexity that arrives on our computer screens through transgenic media as well. Just as our quotidian media have become more personalized, more immediate, and more institutionalized, they have also become more central to our lives. The temporal displacement created through our use of transgenic media, in turn, is refocused through the television and reflected back at us, teaching us how to create stability from the fragments of our own lives.

Our lives are becoming temporally complex. As a monochromic culture, we are obsessed with time, order, and consistency. Yet, as Yu describes at the opening of his *Science Fictional Universe* novel, "Time isn't an orderly stream. Time isn't a placid lake recording each of our ripples. Time is viscous. Time is a massive flow."[45] As the central focus of his book, time is itself both monstrous and tame, both constant and fickle. As the recent spate of temporally complex narratives on television have indicated, our ability to think outside of the normal bounds of temporal linearity allows for greater possibility, both from within ourselves and from our culture at large. For Yu, time is both of these things, and neither: for time is constant change.

Our interaction with transgenic media has forced us to reinvent time. It once took weeks to send a message from one side of our continent to another; now it takes milliseconds to text. This hasn't just created a more communicatively-oriented society; it's also created a new focus on temporality in our own lives. We desire instantaneousness, which makes things like *waiting* and *wanting* all the more pleasurable (and more intolerable). When instant access allows infinite retrieval, the objects that are *not* permanent become even more meaningful.

Time, paradoxically, is perhaps both the most meaningful and the most infinite of all commodities. We never have enough of it, but we seem to have it in spades. From all our wanderings come the most pleasurable experiences: when our sense of time disappears, we are left experiencing the instant. We have become caught in an endless stream of time, from the moment of our birth.

And thus, perhaps I was wrong at the start of this book. Perhaps it is best, after all, not to start with a conclusion, but rather to end this book with that most universal of all beginnings.

Notes

1 Jameson, 27.
2 See "Teen Leisure Habits: TV Still On Top," *Media Literacy Clearinghouse*, 28 Mar 2011. http://www.frankwbaker.com/mediause.htm (accessed 01 Mar 2012).
3 Lisa Parks, "Flexible Microcasting: Gender, Generation, and Television-Internet Convergence," in *Television after TV: Essays on a Medium in Transition*, ed. Lynn Spigel and Jan Olsson (Durham, NC: Duke University Press, 2004), 153.
4 Jostein Gripsrud, "Broadcast Television: The Chances of Its Survival in a Digital Age," in *Television after TV: Essays on a Medium in Transition*, ed. Lynn Spigel and Jan Olsson (Durham, NC: Duke University Press, 2004), 211.
5 See Gillan; Gray, *Show Sold Separately*; Lotz; Ross.
6 William Merrin, "Media Studies 2.0: Upgrading and Open-Sourcing the Discipline," *Interactions: Studies in Communication and Culture* 1, no. 1 (2009): 17–18.
7 Gauntlett, 148.
8 John Hartley, *Uses of Television* (New York: Routledge, 1999).
9 Thorburn and Jenkins, 11.
10 Ibid.
11 Jacques Derrida, *Of Grammatology* (1967), rev. ed., trans. Gayatri Chakravorty Spivak (Baltimore, MD: The Johns Hopkins University Press, 1997).
12 Fiske and Hartley; Ong.
13 Ibid., 2.
14 David Thorburn, "Television as an Aesthetic Medium," 167.
15 Welch.
16 Newcomb, 488.
17 Creeber, *Serial Television*, 7.
18 Marshall McLuhan, *Understanding Media: The Extensions of Man* (1964) (Cambridge, MA: The MIT Press, 1994), 22–33; Scott McCloud, *Understanding Comics: The Invisible Art* (New York: Harper Perennial, 1993).
19 Caldwell, *Televisuality*.
20 Booth, *Digital Fandom*.
21 Creeber, *Serial Television*, 7.
22 Lev Manovich, "What Comes After Remix?" 2007, http://www.manovich.net/DOCS/remix_2007_2.doc (accessed 03 Jan 2008);

see also Laurence Lessig, *Remix: Making Art and Commerce Thrive in the Hybrid Economy* (New York: Penguin, 2008).

23 Raymond Williams, *Cultural and Materialism* (1980) (New York: Verso, 2005), 47.

24 See Johnson.

25 Pierre Bourdieu, *Distinction: A Social Critique of the Judgment of Taste*, trans. Richard Nice (Cambridge, MA: Harvard University Press, 1984).

26 Ibid., 5.

27 Ibid., 71.

28 See SISAPSIS, "Sitting Dance," *YouTube*, 16 Feb 2008, http://www.youtube.com/watch?v=fnFiqRiZh1s (accessed 21 Mar 2011).

29 Leora Broydo, "Attack of the Celebrity Vacuum-Cleaner-Salesman Ghouls," *Salon.com*, 1997, http://www.salon.com/july97/media/media970708.html (accessed 21 Mar 2011).

30 Kelli Marshall, "Gene Kelly, Volkswagon, and Posthumous Performance," *Flow TV*, 2011, http://flowtv.org/2011/04/gene-kelly-volkswagen-andposthumous-performance/ (accessed 08 Apr 2011).

31 Johnson, 9.

32 Jenkins, *Convergence Culture*.

33 Robert Brookey and Paul Booth, "Restricted Play: Synergy and the Limits of Interactivity in *The Lord of the Rings: The Return of the King* Video Game," *Games and Culture* 1, no. 3 (2006); see also Robert Brookey, *Hollywood Gamers: Digital Convergence in the Film and Video Game Industries* (Bloomington, IN: Indiana University Press, 2010).

34 Jay David Bolter, *Writing Space: Computers, Hypertext, and the Remediation of Print*, 2nd ed. (New York: Routledge, 2001).

35 Booth, *Digital Fandom*.

36 Viviane Serfaty, *The Mirror and the Veil: An Overview of American Online Diaries and Blogs* (Amsterdam: Rodopi, 2004).

37 See Steven E. Jones, *The Meaning of Video Games: Gaming and Textual Strategies* (New York: Routledge, 2008), 21.

38 Costas Constandinides, *From Film Adaptation to Post-Celluloid Adaptation: Rethinking the Transition of Popular Narratives and Characters across Old and New Media* (New York: Continuum, 2010).

39 Thomas Elsaesser, "The Mind Game Film," in *Puzzle Films: Complex Storytelling in Contemporary Cinema*, ed. Warren Buckland (Malden, MA: Wiley-Blackwell, 2009); Daly, 85.

40 Marsha Kinder, "Hot Spots, Avatars, and Narrative Fields Forever: Buñuel's Legacy for New Digital Media and Interactive Database Narrative," *Film Quarterly* 55, no. 4 (2002).

41 Kompare, "Publishing Flow."

42 Ibid., 335.

43 See Charlotte Brunsdon, "Bingeing on Box-Sets: The National and the Digital in Television Crime Drama," in *Relocating Television: Television in the Digital Context*, ed. Jostein Gripstrud (New York: Routledge, 2010).

44 Will Brooker, "'Now You're Thinking with Portals': Media Training for a Digital World," *International Journal of Cultural Studies* 13, no. 6 (2010): 554.

Bibliography

Allen, Robert. "Audience-Oriented Criticism and Television." In *Channels of Discourse, Reassembled: Television and Contemporary Criticism*, 2nd ed., edited by Robert Allen, 101–37. Chapel Hill: The University of North Carolina Press, 1992.

———. *Speaking of Soap Operas*. Chapel Hill: University of North Carolina Press, 1985.

Allrath, Gaby. "Life in Dopplegangland: Innovative Character Conception and Alternate Worlds in *Buffy the Vampire Slayer* and *Angel*." In *Narrative Strategies in Television Series*, edited by Gaby Allrath and Marion Gymnich, 132–52. Houndmills, UK: Palgrave Macmillan, 2006.

Altman, Rick. *A Theory of Narrative*. New York: Columbia University Press, 2008.

"Americans Watching More TV Than Ever; Web and Mobile Video Up Too." *Nielsonwire*. 20 May 2009. http://blog.nielsen.com/nielsenwire/online_mobile/americans-watching-more-tv-than-ever/ (accessed 01 Mar 2011).

Ang, Ien. *Living Room Wars: Rethinking Media Audiences for a Postmodern World*. New York: Routledge, 1996.

Arango, Tim. "Broadcast TV Faces Struggle to Stay Viable." *New York Times*. 09 Feb 2009. http://www.nytimes.com/2009/02/28/business/media/28network.html?_r=1&pagallewanted=1&hp (accessed 15 Oct 2010).

Askwith, Ivan. "'Do You Even Know Where This Is Going?' *Lost*'s Viewers and Narrative Premediation." In *Reading Lost: Perspectives on a Hit Television Show*, edited by Roberta Pearson, 159–80. London: I.B. Tauris, 2009.

Atkinson, Cliff. *The Backchannel: How Audiences Are Using Twitter and Social Media and Changing Presentations Forever*. Berkeley, CA: New Riders, 2010.

Bal, Mieke. *Narratology: Introduction to the Theory of Narrative*. 3rd ed. Translated by Christine Van Boheemen. Toronto, ON: University of Toronto Press, 2009.

———. *On Story-Telling: Essays in Narratology*. Sonoma, CA: Polebridge Press, 1991.

Barthes, Roland. *S/Z*. 1970. Translated by Richard Miller. New York: Farrar, Straus and Girous, 1974.

Baudrillard, Jean. "The Ecstasy of Communication." In *The Anti-Aesthetic*, edited by Hal Foster, 126–34. Port Townsend, WA: Bay Press, 1983.

———. *For a Critique of the Political Economy of the Sign*. 1968. Translated by Charles Levin. NY: Telos Press, 1981.

———. *The Illusion of the End*. Translated by Chris Turner. Stanford, CA: Stanford University Press, 1992.

Baym, Nancy. *Personal Connections in the Digital Age*. Cambridge, UK: Polity Press, 2010.

Bays, Carter, and Craig Thomas, writers. "Wait for It." *How I Met Your Mother*. Los Angeles, CA: CBS, 2007.

Beer, David, and Roger Burrows. "Sociology and, of and in Web 2.0: Some Initial Considerations." *Sociological Research Online* 12, no. 5 (2007): http://www.socresonline.org.uk/12/5/17.html (accessed 15 Aug 2011).

Benkler, Yochai. *The Wealth of Networks*. New Haven, CT: Yale University Press, 2006.

Best, Kirsty. "Living in the Control Society: Surveillance, Users and Digital Screen Technologies." *International Journal of Cultural Studies* 13, no. 1 (2010): 5–24.

Boddy, William. "Interactive Television and Advertising Form in Contemporary U.S. Television." In *Television after TV: Essays on a Medium in Transition*, edited by Lynn Spigel and Jan Olsson, 113–32. Durham, NC: Duke University Press, 2004.

———. "Redefining the Home Screen: Technological Convergence as Trauma and Business Plan." In *Rethinking Media Change: The Aesthetics of*

Transition, edited by David Thorburn and Henry Jenkins, 191–200. Cambridge, MA: The MIT Press, 2004.

Bolter, Jay David. *Writing Space: Computers, Hypertext, and the Remediation of Print*. 2nd ed. New York: Routledge, 2001.

Bolter, Jay David, and Richard Grusin. *Remediation: Understanding New Media*. Cambridge, MA: The MIT Press, 1999.

Booth, Paul. *Digital Fandom*. New York: Peter Lang, 2010.

———. "Frak-tured Postmodern Lives, or How I Found Out I Was a Cylon." In *Battlestar Galactica and Philosophy*, edited by Josef Steiff and Tristan Tamblin, 17–28. Peru, IL: Open Court Publishing, 2008.

———. "Narractivity and the Narrative Database: Media-Based Wikis as Interactive Fan Fiction." *Narrative Inquiry* 19, no. 2 (2009): 372–92.

Booth, Wayne. *The Rhetoric of Fiction*. Chicago, IL: University of Chicago Press, 1961.

Bordwell, David. *Narration in the Fiction Film*. Madison: University of Wisconsin Press, 1985.

Bourdieu, Pierre. *Distinction: A Social Critique of the Judgment of Taste*. 1979. Translated by Richard Nice. Cambridge, MA: Harvard University Press, 1984.

boyd, danah. "Social Network Sites as Networked Publics: Affordances, Dynamics, and Implications." In *A Networked Self: Identity, Community, and Culture on Social Network Sites*, edited by Zizi Papacharissi, 37–58. New York: Routledge, 2011.

———. "Why Youth (Heart) Social Network Sites: The Role of Networked Publics in Teenage Social Life." In *Youth, Identity, and Digital Media Volume*, edited by David Buckingham, 119–42. Cambridge, MA: The MIT Press, 2007. http://www.danah.org/papers/ WhyYouthHeart.pdf (accessed 10 Jan 2008).

boyd, danah, and Nicole Ellison. "Social Network Sites: Definition, History, and Scholarship." *Journal of Computer-Mediated Communication* 13, no. 1 (2007): http://jcmc.indiana.edu/vol13/issue1/boyd.ellison.html (accessed 01 Jan 2010).

Brinkmann, Martin. "Six Great Google Maps Mashups." *ghacks.com*. 08 Nov 2007. http://www.ghacks.net/2007/11/08/six-great-google-maps-mashups/ (accessed 23 Dec 2010).

Brooker, Will. "'Now You're Thinking with Portals': Media Training for a Digital World." *International Journal of Cultural Studies* 13, no. 6 (2010): 553–73.

Brookey, Robert. *Hollywood Gamers: Digital Convergence in the Film and Video Game Industries*. Bloomington: Indiana University Press, 2010.

Brookey, Robert, and Paul Booth. "Restricted Play: Synergy and the Limits of Interactivity in *The Lord of the Rings: The Return of the King* Video Game." *Games and Culture* 1, no. 3 (2006): 214–30.

Broydo, Leora. "Attack of the Celebrity Vacuum-Cleaner-Salesman Ghouls." *Salon.com*. 1997. http://www.salon.com/july97/media/media970708.html (accessed 21 Mar 2011).

Bruno, Giuliana. "Ramble City: Postmodernism and *Blade Runner*." *October* 41 (1987): 61–74.

Bruns, Axel. *Blogs, Wikipedia, Second Life, and Beyond: From Production to Produsage*. New York: Peter Lang, 2008.

Brunsdon, Charlotte. "Bingeing on Box-Sets: The National and the Digital in Television Crime Drama." In *Relocating Television: Television in the Digital Context*, edited by Jostein Gripstrud, 63–75. New York: Routledge, 2010.

Bukatman, Scott. *Terminal Identity*. Durham, NC: Duke University Press, 1993.

Bush, Vannevar. "As We May Think." *Atlantic Monthly*. July 1945. http://www.theatlantic.com/magazine/print/1969/12/as-we-may-think/3881/ (accessed 15 Mar 2011).

Caldwell, John. "Convergence Television: Aggregating Form and Repurposing Content in the Culture of Conglomeration." In *Television after TV*, edited by Lynn Spigel and Jan Olsson, 41–74. Durham, NC: Duke University Press, 2004.

———. *Televisuality: Style, Crisis, and Authority in American Television*. Piscataway, NJ: Rutgers University Press, 1995.

Castells, Manuel. *The Internet Galaxy*. New York: Oxford University Press, 2003.

———. *The Rise of the Network Society: The Information Age: Economy, Society, and Culture Volume 1*. 2nd ed. Malden, MA: Wiley-Blackwell, 2009.

Chamberlain, Daniel. "Media Interfaces, Networked Media Space, and the Mass Customization of Everyday Space." In *Flow TV: Television in the Age of Media Convergence*, edited by Michael Kackman, Marnie Binfield, Matthew Thomas Payne, Allison Perlman, Bryan Sebok, 13–29. New York: Routledge, 2010.

"Character Connection." *Lostpedia*. 2011. http://lostpedia.wikia.com/wiki/Character_connections (accessed 04 Mar 2011).

Charles, Alec. "Ideology of Anachronism: Television, History, and the Nature of Time." In *Time and Relative Dissertations in Space: Critical Perspectives on Doctor Who*, edited by David Butler, 108–22. Manchester, UK: Manchester University Press, 2007.

Chatman, Seymour. *Coming to Terms*. Ithaca, NY: Cornell University Press, 1990.

———. *Story and Discourse: Narrative Structure in Fiction and Film*. Ithaca, NY: Cornell University Press, 1978.

———. "What Novels Can Do That Films Can't (And Vice Versa)." *Critical Inquiry* 7, no. 1 (1980): 121–40.

Clarke, M. J. "Lost and Mastermind Narration." *Television and New Media* 11, no. 2 (2010): 123–142.

Cohen, Jonathan. "Defining Identification: A Theoretical Look at the Identification of Audiences with Media Characters." *Mass Communication and Society* 4, no. 3 (2001): 245–26.

Confino, Alon. "Collective Memory and Cultural History: Problems of Method." *The American Historical Review* 102, no. 5 (1997): 1386–403.

Constandinides, Costas. *From Film Adaptation to Post-Celluloid Adaptation: Rethinking the Transition of Popular Narratives and Characters across Old and New Media*. New York: Continuum, 2010.

Cook, Pam. *Screening the Past: Memory and Nostalgia in Cinema*. New York: Routledge, 2005.

Cooke, Lynn. "A Visual Convergence of Print, Television, and The Internet: Charting 40 Years of Design Change in News Presentation." *New Media and Society* 7, no. 1 (2005): 22–46.

Cowan, Douglas E. *Sacred Space: The Quest for Transcendence in Science Fiction Film and Television*. Waco, TX: Baylor University Press, 2010.

Craig, Susanne, and Andrew Ross Sorkin. "Goldman Offering Clients a Chance to Invest in Facebook." 02 Jan 2011. http://dealbook.nytimes.com/2011/01/02/goldman-invests-in-facebook-at-50-billion-valuation/ (accessed 02 Jan 2011).

Crary, Jonathan. "Spectacle, Attention, Counter-Memory." *October* 50 (1989): 96–107.

Creeber, Glen. "*Heroes.*" In *The Television Genre Book,* 2nd ed., edited by Glen Cleeber, Toby Miller, and John Tulloch, 57. Houndmills, UK: Palgrave Macmillan, 2008.

———. *Serial Television: Big Drama on the Small Screen.* London: British Film Institute, 2008.

Currie, Gregory. *Narrative and Narrators: A Philosophy of Stories*. Oxford, UK: Oxford University Press, 2010.

Curtis, Richard, writer. "Vincent and the Doctor." *Doctor Who*. London: BBC, 2010.

Daly, Kristen. "Cinema 3.0: The Interactive-Image." *Cinema Journal* 50, no. 1 (2010): 81–98.

Dancyger, Ken. *The Technique of Film and Video Editing*. 3rd ed. New York: Focal Press, 2002.

Davisson, Amber. "Beyond the Borders of Red and Blue States: Google Maps as a Site of Rhetorical Experimentation in the 2008 Presidential Election." *Journal of Rhetoric and Public Affairs* 11, no. 4 (2011): 101–23.

de Certeau, Michel. *The Practice of Everyday Life*. 1980. Translated by Steven Randall. Berkeley: University of California Press, 1984.

Debord, Guy. *The Society of the Spectacle*. 1967. Translated by D. Nicholson-Smith. Detroit: Black & Red, 1983.

Dena, Christy. "Emerging Participatory Culture Practices: Player-Created Tiers in Alternate Reality Games." *Convergence: The International Journal of Research into New Media Technologies* 14, no. 1 (2008): 41–57.

Derrida, Jacques. *Of Grammatology*. 1967. Rev. ed. Translated by Gayatri Chakravorty Spivak. Baltimore, MD: The Johns Hopkins University Press, 1997.

"Desmond's Photograph." *Lostpedia*. http://lostpedia.wikia.com/wiki/Desmond's_photograph (accessed 01 Feb 2011).

Dibbell, Julian. "A Rape in Cyberspace: How an Evil Clown, a Haitian Trickster Spirit, Two Wizards, and a Cast of Dozens Turned a Database into a Society." *The Village Voice*. 23 Dec 1993. http://www.juliandibbell.com/texts/bungle_vv.html (accessed 14 Aug 2011).

Dicks, Terrance. *Day of the Daleks*. London: Target Books, 1974.

DiPaolo, Marc Edward. "Political Satire and British-American Relations in Five Decades of *Doctor Who*." *Journal of Popular Culture* 43, no. 5 (2010): 964–87.

Donath, Judith, and danah boyd. "Public Displays of Connection." *BT Technology Journal* 22, no. 4 (2004): 71–82.

Draoulec, Anne, and Marie-Paule Péry-Woodley. "Time Travel in Text: Temporal Framing in Narratives and Non-Narratives." In *Determination of Information and Tenor in Texts, Proceedings of Multidisciplinary Approaches to Discourse*, edited by L. Lagerwerf, W. Spooren & L.

Degand, 133–42. Amsterdam: Stichting Neerlandistiek & Münster, Nodus Publikationen, 2003. http://w3.erss.univtlse2.fr/textes/pagespersos/pery/articles/MAD03.pdf (accessed 01 Apr 2011).

Eco, Umberto. "On the Ontology of Fictional Characters: A Semiotic Approach." *Sign Systems Studies* 37, no. 1/2 (2009): 82–98.

Eichler, Alex. "Meet Keith Urbahn, First to Tweet News of Bin Laden." *The Atlantic Wire*. 04 May 2011. http://www.theatlanticwire.com/national/2011/05/meet-keith-urbahn-first-tweet-news-bin-laden/37331 (accessed 29 May 2011).

Elsaesser, Thomas. "The Mind Game Film." In *Puzzle Films: Complex Storytelling in Contemporary Cinema*, edited by Warren Buckland, 13–41. Malden, MA: Wiley-Blackwell, 2009.

Espenson, Jane. "Playing Hard to 'Get'—How to Write Cult TV." In *The Cult TV Book*, edited by Stacey Abbott, 44–54. London, UK: I. B. Taurus, 2010.

Eternal Darkness: Sanity's Requiem. St. Catharines, CA: Silicon Knights, 2002.

Fiske, John. "The Cultural Economy of Fandom." In *The Adoring Audience*, edited by Lisa A. Lewis, 37–42. London: Routledge, 1992.

Fiske, John, and John Hartley. *Reading Television*. 2nd ed. New York: Routledge, 2003.

Fletcher, Dan. "What Happens to Your Facebook After You Die?" *Time*. 28 Oct 2009. http://www.time.com/time/business/article/0,8599,1932803,00.html (accessed 5 May 2011).

Foucault, Michel. "Self-Writing." 1983. In *Ethics: Subjectivity and Truth,* Vol 1., edited by Paul Rabinow, translated by Robert Hurley and Others, 207–22. New York: The New Press, 1997.

———. "Technologies of the Self." 1982. In *Ethics: Subjectivity and Truth,* Vol 1, edited by Paul Rabinow, Translated by Robert Hurley and Others, 223–52. New York: The New Press, 1997.

Friedberg, Anne. "The Virtual Window." In *Rethinking Media Change: The Aesthetics of Transition*, edited by David Thorburn and Henry Jenkins, 337–54. Cambridge, MA: The MIT Press, 2004.

Gauntlett, David. "Media Studies 2.0: A Response." *Interactions: Studies in Communication and Culture* 1, no. 1 (2009): 147–57.

Gendler, Jason. "*Primer*: The Perils and Paradoxes of Restricted Time Travel Narration." *Nebula* 3, no. 4 (2009): http://www.iiav.nl/ezines/web/Nebula/ 2009/No2/Gendler.pdf (accessed June 2010).

Genette, Gérard. *Narrative Discourse: An Essay in Method*. 1972. Translated by Jane Lewin. Ithaca, NY: Cornell University Press, 1980.

———. *Narrative Discourse Revisited.* 1983. Translated by Jane Lewin. Ithaca, NY: Cornell University Press, 1988.

Gillan, Jennifer. *Television and New Media: Must Click TV.* New York: Routledge, 2010.

Gitlin, Todd. "'Make It Look Messy: Striking a Note with *Hill Street Blues.*" *American Film* (Sept 1981): 45–9, 76.

Goffman, Erving. *The Presentation of Self in Everyday Life.* New York: Doubleday, 1959.

Goldstein, Laurence. "A Unified Solution to Some Paradoxes." *Proceedings of the Aristotelian Society* 100 (2000): 53–74.

Gorman, Andrew. "*Back to the Future:* Oedipus as Time Traveller." In *Liquid Metal: A Science Fiction Reader,* edited by Sean Redmond, 116–25. Middlesex, UK: Wallflower Press, 2004.

Gorman, David. "Reconsidering Narrative Theory Rhetorically." *Poetics Today* 31 (2010): 353–58.

Gray, Jonathan. *Show Sold Separately.* New York: New York University Press, 2010.

———. *Television Entertainment.* New York: Routledge, 2008.

Gripsrud, Jostein. "Broadcast Television: The Chances of Its Survival in a Digital Age." In *Television after TV: Essays on a Medium in Transition,* edited by Lynn Spigel and Jan Olsson, 210–23. Durham, NC: Duke University Press, 2004.

Grossman, Lev. "*Time*'s Person of the Year: You." *Time.* 2006. http://www.time.com/time/magazine/article/0,9171,1569514,00.html (accessed 18 Sept 2008).

Gunkel, David. *Thinking Otherwise.* West Lafayette, IN: Purdue University Press, 2009.

Gunkel, David J., and Ann Hetzel Gunkel. "Terra Nova 2.0—The New Worlds of MMORPGs." *Critical Studies in Media Communication* 26, no. 2 (2009): 104–27.

Gunning, Tom. "Re-Newing Old Technologies: Astonishment, Second Nature, and the Uncanny in Technology from the Previous Turn-of-the-Century." In *Rethinking Media Change: The Aesthetics of Transition,* edited by David Thorburn and Henry Jenkins, 39–60. Cambridge, MA: The MIT Press, 2004.

Gurney, David. "'It's Just Like a Mini-Mall!' Textuality and Participatory Culture on YouTube." In *Flow TV: Television in the Age of Media Convergence,* edited by Michael Kackman, Marnie Binfield, Matthew

Thomas Payne, Allison Perlman, and Bryan Sebok, 30–45. New York: Routledge, 2010.

Gwenllian-Jones, Sara. "The Sex Lives of Cult Television Characters." *Screen* 43, no. 1 (2002): 79–90.

———. "Virtual Reality and Cult Television." In *Cult Television*, edited by Sara Gwenllian-Jones and Roberta E. Pearson, 83–98. Minneapolis: University of Minnesota Press, 2004.

Halavais, Alex. *Search Engine Society*. Cambridge, UK: Polity Press, 2009.

Hanley, Richard. *Is Data Human? The Metaphysics of Star Trek*. New York: Basic Books, 1997.

Hardy, Jonathan. "Mapping Commercial Intertextuality: HBO's *True Blood*." *Convergence: The International Journal of Research into New Media Technologies* 17, no. 1 (2011): 7–17.

Hartley, John. *Uses of Television*. New York: Routledge, 1999.

Haskins, Ekaterina. "Between Archive and Participation: Public Memory in a Digital Age." *Rhetoric Society Quarterly* 37, no. 4 (2007): 401–22.

Hassan, Robert. *The Information Society*. Cambridge, UK: Polity Press, 2008.

Haynes, Holly. "As *Lost* Ends, Creators Explain How They Did It, What's Going on." *Wired* 18, no. 5 (19 April 2010). http://www.wired.com/magazine/2010/04/ff_lost/5/ (accessed 14 Mar 2011).

Hellekson, Karen. "Memory and the 1996 American Remake of *Doctor Who*." In *American Remakes of British Television: Transformations and Mistranslations*, edited by Carlen Lavigne and Heather Marcovitch, 159–72. Lanham, MD: Lexington Books, 2011.

Herman, David. *Story Logic: Problems and Possibilities of Narrative*. Lincoln: Univeristy of Nebraska Press, 2002.

Hills, Matt. "Absent Epic, Implied Story Arcs, and Variation on a Narrative Theme: *Doctor Who* (2005–2008) as Cult/Mainstream Television." In *Third Person: Authoring and Exploring Vast Narratives*, edited by Pat Harrigan and Noah Wardrip-Fruin, 333–43. Cambridge, MA: The MIT Press, 2009.

———. *Fan Cultures*. London: Routledge, 2002.

———. *Triumph of a Time Lord: Regenerating Doctor Who in the Twenty-First Century*. London, UK: Taurus, 2010.

Holdsworth, Amy. "Television Resurrections: Television and Memory." *Cinema Journal* 47, no. 3 (2008): 137–44.

Holmes, Robert, writer. "The Ribos Operation." *Doctor Who*. London: BBC, 1978.

Horton, David, and Richard Wohl. "Mass Communication and Parasocial Interaction." *Psychiatry* 19 (1956): 215–29.

Hutton, Patrick. "Collective Memory and Collective Mentalities." *Historical Relfections* 15, no. 2 (1988): 311–22.

Igartua, Juan-José. "Identification with Characters and Narrative Persuasion through Fictional Feature Films." *Communications* 35, no. 4 (2010): 347–73.

Illich, Ivan, and Barry Sanders. *ABC: The Alphabetization of the Popular Mind.* San Francisco: North Point Press, 1988.

Internet World Stats: Usage and Population Statistics. "Internet Users in the World." Mar 2011. http://www.internetworldstats.com/stats.htm (accessed 16 Aug 2011).

Jameson, Fredric. *Postmodernism: Or, the Cultural Logic of Late Capitalism.* Durham, NC: Duke University Press, 1991.

Jenkins, Henry. *Convergence Culture: Where Old and New Media Collide*. New York: New York University Press, 2006.

———. *Textual Poachers*. New York: Routledge, 1992.

———. "Transmedia Storytelling." *Technology Review.* 15 Jan 2003. http://www.technologyreview.com/biotech/13052/?a=f (accessed 29 Dec 2010).

———. *The Wow Climax: Tracing the Emotional Impact of Popular Culture*. New York: New York University Press, 2007.

Jensen, Joli. "Fandom as Pathology: The Consequences of Characterization." In *The Adoring Audience*, edited by Lisa A. Lewis, 9-29. New York: Routledge, 1992.

Johnson, Steven. *Everything Bad Is Good for You*. New York: Riverhead Books, 2005.

Jones, Glynn, writer. "The Space Museum." *Doctor Who*. London: BBC, 1965.

Jones, Steven E. *The Meaning of Video Games: Gaming and Textual Strategies.* New York: Routledge, 2008.

Kackman, Michael, Marnie Binfield, Matthew Thomas Payne, Allison Perlman, and Bryan Sebok, ed. *Flow TV: Television in the Age of Media Convergence*. New York: Routledge, 2011.

Kilburn, Matthew. "Bargains of Necessity? *Doctor Who,* Culloden and Fictionalising History at the BBC in the 1960s." In *Time and Relative Dissertations in Space: Critical Perspectives on Doctor Who*, edited by David Butler, 68–85. Manchester, UK: Manchester University Press, 2007.

Kinder, Marsha. "Hot Spots, Avatars, and Narrative Fields Forever: Buñuel's Legacy for New Digital Media and Interactive Database Narrative." *Film Quarterly* 55, no. 4 (2002): 2–15.

Kogan, Lauren. "Once or Twice upon a Time: Temporal Simultaneity and the *Lost* Phenomenon." *Film International* 4, no. 2 (2006): 44–55.

Kompare, Derek. "Publishing Flow: DVD Box Sets and the Reconception of Television." *Television and New Media* 7, no. 4 (2006): 335–60.

Konieczny, Piotr. "Governance, Organization, and Democracy on the Internet: The Iron Law and the Evolution of Wikipedia." *Sociological Forum* 24, no. 1 (2009): 162–92.

Kozloff, Sarah. "Narrative Theory and Television." In *Channels of Discourse, Reassembled: Television and Contemporary Criticism,* 2nd ed., edited by Robert C. Allen, 67–100. Chapel Hill: The University of North Carolina Press, 1992.

Landow, George. *Hypertext 3.0: Critical Theory and New Media in an Era of Globalization*. Baltimore, MD: The Johns Hopkins University Press, 2006.

Lavery, David. "*Lost* and Long-Term Television Narrative." In *Third Person: Authoring and Exploring Vast Narratives,* edited by Pat Harrigan and Noah Wardrip-Fruin, 313–22. Cambridge, MA: The MIT Press, 2009.

Lavik, Erlend. "Forward to the Past: The Strange Case of *The Wire*." In *Relocating Television: Television in the Digital Context,* edited by Jostein Gripstrud, 76–87. New York: Routledge, 2010.

Lem, Stanislaw. "The Time Travel Story and Related Matters of Science-Fiction Structuring." *Microworlds*. London: Mandarin, 1991.

Lenhart, Amanda, Kristen Purcell, Aaron Smith, and Kathryn Zickuhr. "Social Media and Mobile Internet Use Among Teens and Young Adults." *Pew Research Center*. 2010. http://pewresearch.org/pubs/1484/social-media-mobile-internet-use-teens-millennials-fewer-blog (accessed 10 Mar 2010).

Lessig, Laurence. *Remix: Making Art and Commerce Thrive in the Hybrid Economy*. New York: Penguin, 2008.

Levinson, Paul. *New New Media*. New York: Allyn & Bacon, 2009.

Lévy, Pierre. *Collective Intelligence: Mankind's Emerging World in Cyberspace*. 1994. Translated by Robert Bononno. New York: Perseus Pub, 1997.

Lewis, David. "The Paradoxes of Time Travel." *American Philosophical Quarterly* 13, no. 2 (1976): 145–52.

Littau, Karin. "Media, Mythology and Morphogenesis: *Aliens*™." *Convergence: The International Journal of Research into New Media Technologies* 17, no. 1 (2011): 19–36.

"Lost Connections." *Lost* Season 2 DVD. Los Angeles: Buena Vista Home Entertainment, 2006.

Lotz, Amanda. *The Television Will Be Revolutionized*. New York: New York University Press, 2007.

Madden, Mary, and Aaron Smith. "Reputation Management and Social Media." *Pew Research Center*. 2010. http://www.pewinternet.org/Reports/2010/Reputation-Management.aspx (accessed 01 June 2010).

Manovich, Lev. *The Language of New Media*. Cambridge, MA: The MIT Press, 2001.

———. "What Comes After Remix?" 2007. http://www.manovich.net/DOCS/remix_2007_2.doc (accessed 03 Jan 2008).

Marcovitch, Heather. "Memories of Mars: *Life on Mars* and the Discursive Practices of Memory." In *American Remakes of British Television: Transformations and Mistranslations*, edited by Carlen Lavigne and Heather Marcovitch, 173–92. Lanham, MD: Lexington Books, 2011.

Marshall, Kelli. "Gene Kelly, Volkswagon, and Posthumous Performance." *Flow TV*. 08 Apr 2011. http://flowtv.org/2011/04/gene-kelly-volkswagen-andposthumous-performance/ (accessed 08 Apr 2011).

McCloud, Scott. *Understanding Comics: The Invisible Art*. New York: Harper Perennial, 1993.

McLuhan, Marshall. *Understanding Media: The Extensions of Man*. Cambridge, MA: The MIT Press, 1999.

Meadows, Mark Stephen. *Pause and Effect*. Berkeley, CA: New Riders, 2003.

Menace, Delton. "Let's Keep Rumours Clean." *Tardis.wikia*. 2010. http://tardis.wikia.com/wiki/Talk:The_Big_Bang (accessed 18 June 2010).

Merrin, William. "Media Studies 2.0: Upgrading and Open-Sourcing the Discipline." *Interactions: Studies in Communication and Culture* 1, no. 1 (2009): 17–35.

Mittell, Jason. "*Lost* in a Great Story: Evaluation in Narrative Television (and Television Studies)." In *Reading Lost: Perspectives on a Hit Television Show*, edited by Roberta Pearson, 119–38. London: I.B. Tauris, 2009.

———. "Narrative Complexity in Contemporary American Television." *The Velvet Light Trap* 58 (2006): 29–40.

———. "Previously On: Prime Time Serials and the Mechanics of Memory." In *Intermediality and Storytelling,* edited by Marina Grishakova and Marie-Laure Ryan, 78–98. Boston, MA: De Gruyter, 2010.

———. "Previously On: Prime Time Serials and the Mechanics of Memory." *Just TV.* 03 July 2009. http://justtv.wordpress.com/2009/07/03/previously-on-prime-time-serials-and-the-mechanics-of-memory/ (accessed 03 Mar 2010).

———. "Sites of Participation: Wiki Fandom and the Case of *Lostpedia.*" *Transformative Works and Cultures* 3 (2009): http://journal.transformativeworks.org/index.php/twc/article/view/118/117 (accessed 01 Jan 2010).

———. *Television and American Culture.* New York: Oxford University Press, 2010.

———. "TiVoing Childhood: Time-Shifting a Generation's Concept of Television." In *Flow TV: Television in the Age of Media Convergence,* edited by Michael Kackman, Marnie Binfield, Matthew Thomas Payne, Allison Perlman, and Bryan Sebok, 46–54. New York: Routledge, 2010.

———. "Trapped in a Web of Meta-Wikipedia." *Just TV.* 06 May 2009. http://justtv.wordpress.com/2009/05/06/trapped-in-a-web-of-meta-wikipedia/ (accessed 06 May 2009).

Modleski, Tania. "The Rhythms of Reception: Daytime Television and Women's Work." In *Regarding Television,* edited by E. Ann Kaplan, 67–75. Fredricksburg, MD: The American Film Institute, 1983.

Moffat, Steven, writer. "The Beast Below." *Doctor Who.* London: BBC, 2010.

———. "The Big Bang." *Doctor Who.* London: BBC, 2010.

———. "Blink." *Doctor Who.* London: BBC, 2007.

———. "The Pandorica Opens." *Doctor Who.* London: BBC, 2010.

———. "Silence in the Library." *Doctor Who.* London: BBC, 2008.

Moran, James, writer. "The Fires of Pompeii." *Doctor Who.* London: BBC, 2009.

Morgasen, David. "Obama and McCain – Dance Off!" *YouTube.* 23 Oct 2008. http://www.youtube.com/watch?v=wzyT9-9lUyEzz (accessed 01 Mar 2011).

Murray, Janet. *Hamlet on the Holodeck: The Future of Narrative in Cyberspace.* Cambridge, MA: The MIT Press, 1997.

"MySpace." *Wikipedia.* 2011. http://en.wikipedia.org/wiki/Myspace (accessed 16 March 2011).

Nahin, Paul. *Time Machines: Time Travel in Physics, Metaphysics, and Science Fiction*. New York: Springer, 1993.

Naremore, James. "Authorship and the Cultural Politics of Film Criticism." *Film Quarterly* 44, no. 1 (1990): 14–23.

Nation, Terry, writer. "Genesis of the Daleks." *Doctor Who*. London: BBC, 1975.

Newcomb, Horace. "Towards a Television Aesthetic." In *Television: The Critical View,* 3rd ed., edited by Horace Newcomb, 478–94. New York: Oxford University Press, 1982.

Newman, Ira. "Virtual People: Fictional Characters through the Frames of Reality." *The Journal of Aesthetics and Art Criticism* 67 (2009): 73–82.

Nikolajeva, Maria. *The Rhetoric of Character in Children's Literature*. Lanham, MD: The Scarecrow Press, 2002.

Nora, Pierre. "Between Memory and History: Les Lieux de Mèmoire." *Representations* 26 (1989): 7–24.

———, ed. *Les Lieux de Mèmoire, Vol 7*. Paris: Les Frances, La Republique, Le Nation, 1992.

O'Neil, Mathieu. "Wikipedia: Experts Are Us." *Le Monde Diplomatique English Edition*. 15 May 2009. http://mondediplo.com/2009/05/15wikipedia (accessed 01 June 2010).

Olick, Jeffrey K., and Joyce Robbins. "Social Memory Studies: From 'Collective Memory' to the Historical Sociology of Mnemonic Practices." *Annual Review of Sociology* 24 (1998): 105–40.

Ong, Walter. *Orality and Literacy*. London: Routledge, 2002.

Örnebring, Henrik. "Alternate Reality Gaming and Convergence Culture: The Case of *Alias*." *International Journal of Cultural Studies* 10, no. 4 (2007): 445–62.

Page, Adrian. "Postmodern Drama." In *The Television Genre Book,* 2nd ed., edited by Glen Cleeber, Toby Miller, and John Tulloch, 43–6. UK: Palgrave Macmillan, 2008.

Palfrey, John, and Urs Gasser. *Born Digital: Understanding the First Generation of Digital Natives*. New York: Basic Books, 2008.

Papacharissi, Zizi. *A Private Sphere: Democracy in the Digital Age*. Cambridge, UK: Polity Press, 2010.

———. "Conclusion: A Networked Self." In *A Networked Self: Identity, Community, and Culture on Social Network Sites,* edited by Zizi Papacharissi, 304–18. New York: Routledge, 2011.

———, ed. *A Networked Self: Identity, Community, and Culture on Social Network Sites*. New York: Routledge, 2011.

Parkin, Lance. *AHistory: An Unauthorized History of the Doctor Who Universe*. Des Moines, IA: Mad Norwegian Press, 2006.

———. "Truths Universally Acknowledged: How the 'Rules' of *Doctor Who* Affect the Writing." In *Third Person: Authoring and Exploring Vast Narratives*, edited by Pat Harrigan and Noah Wardrip-Fruin, 13–24. Cambridge, MA: The MIT Press, 2009.

Parks, Lisa. "Flexible Microcasting: Gender, Generation, and Television-Internet Convergence." In *Television after TV: Essays on a Medium in Transition*, edited by Lynn Spigel and Jan Olsson, 133–56. Durham, NC: Duke University Press, 2004.

Parks, Malcolm R. "Social Network Sites as Virtual Communities." In *A Networked Self: Identity, Community, and Culture on Social Network Sites*, edited by Zizi Papacharissi, 105–23. New York: Routledge, 2011.

Pearson, Lars. *I, Who 3: The Unauthorized Guide to Doctor Who Novels and Audios*. Des Moines, IA: Mad Norwegian Press, 2003.

Pearson, Roberta. "Anatomising Gilbert Grissom: The Structure and Function of the Televisual Character." In *Reading CSI: Television under the Microscope*, edited by Michael Allen, 39–59. London: I.B. Tauris, 2007.

———. "Chain of Events: Regimes of Evaluation and *Lost*'s Construction of the Televisual Character." In *Reading Lost: Perspectives on a Hit Television Show*, edited by Roberta Pearson, 139–59. London: I.B. Tauris, 2009.

———. "*Lost* in Translation: From Post-Network to Post-Television." In *Quality: Contemporary American Television and Beyond*, edited by Kim Akass and Janet McCabe, 239–56. London: I.B. Tauris, 2007.

Phelan, James. *Living to Tell about It: A Rhetoric and Ethics of Character Narration*. Ithaca, NY: Cornell University Press, 2005.

———. *Reading People, Reading Plots: Character, Progression, and the Interpretation of Narrative*. Chicago, IL: University of Chicago Press, 1989.

Pinkner, Jeff, and Joel Wyman, writers. "The Day We Died." *Fringe*. Los Angeles, CA: Fox, 2011.

Plato. *Phaedrus*. Translated by W. C. Helmbold and W. G. Rabinowitz. Indianapolis: Library of Liberal Arts-Bobbs, 1956.

Pogue, David. "Some E-Books Are More Equal than Others." *New York Times online*, 17 July 2009, http://pogue.blogs.nytimes.com/2009/07/17/some-e-books-are-more-equal-than-others/ (accessed 10 Apr 2011).

Porter, Lynnette, and David Lavery. Lost's *Buried Treasures,* 2nd ed. Naperville, IL: Sourcebooks, Inc., 2010.

Press, Andrea L, and Bruce A. Williams. *The New Media Environment: An Introduction.* Malden, MA: Wiley-Blackwell, 2010.

"Press Room." *Facebook.* 2011. http://www.facebook.com/press/info.php?statistics (accessed 16 Mar 2011).

Redmond, Sean, ed. *Liquid Metal: A Science Fiction Reader*. Middlesex, UK: Wallflower Press, 2004.

"Regeneration (episode)." *Memory Alpha.* 2010. http://memory-alpha.org/wiki/Regeneration_(episode) (accessed 22 Dec 2010).

Rhodes, Bradley J. "The Wearable Remembrance Agent: A System for Augmented Memory." *Personal Technologies* 1 (1997): 218–24.

Rickman, Greg. *Science Fiction Film Reader.* New York: Limelight, 2004.

Ricoeur, Paul. "Narrative Time." *Critical Inquiry* 7, no. 1 (1980): 169–90.

Robertson, Stephen. "What's Wrong with Online Readings? Text, Hypertext, and the History Web." *The History Teacher* 39, no. 4 (2006): http://www.historycooperative.org/journals/ht/39.4/robertson.html.

Rosenstone, Robert A. *History in Images/History in Words: Reflections on the Possibility of Really Putting History onto Film*. Cambridge, MA: Harvard University Press, 1995.

Rosenzweig, Roy. "Can History Be Open Source? Wikipedia and the Future of the Past." *Journal of American History* 93, no. 1 (2006): 117–46.

Roser, Nancy, Miriam Martinez, Charles Fuhrken, and Kathleen McDonnold. "Characters as Guides to Meaning." *The Reading Teacher* 60, no. 6 (2007): 548–59.

Ross, Sharon Marie. *Beyond the Box: Television and the Internet*. Malden, MA: Blackwell, 2008.

Ryan, Marie-Laure. *Narrative as Virtual Reality: Immersion and Interactivity in Literature and Electronic Media*. Baltimore, MD: The Johns Hopkins University Press, 2001.

Sandvoss, Cornel. *Fans: The Mirror of Consumption*. Malden, MA: Polity Press, 2005.

Schapker, Alison, and Monica Owusu-Breen. Writers. "Marionette." *Fringe* Los Angeles, CA: Fox, 2010.

Sconce, Jeffrey. *Haunted Media: Electronic Presence from Telegraphy to Television*. Durham, NC: Duke University Press, 2000.

———. "What If? Charting Television's New Textual Boundaries." In *Television after TV*, edited by Lynn Spigel and Jan Olsson, 93–113. Durham, NC: Duke University Press, 2004.

Seiter, Ellen. *Television and New Media Audiences*. Oxford: Oxford University Press, 1999.

Selter, Brian. "TV Viewing Continues to Edge Up." *New York Times,* 02 Jan 2011.

Serfaty, Viviane. *The Mirror and the Veil: An Overview of American Online Diaries and Blogs.* Amsterdam: Rodopi, 2004.

Shearman, Robert. "Defending the Museum." Documentary on DVD. *Doctor Who: The Space Museum*. London: BBC, 2010.

Shearman, Robert, and Toby Hadoke. *Running Through Corridors: Rob and Toby's Marathon Watch of Doctor Who.* Des Moines, IA: Mad Norwegian Press, 2010.

Shimpach, Shawn. *Television in Transition: The Life and Afterlife of the Narrative Action Hero*. Oxford, UK: Wiley-Blackwell, 2010.

Shirky, Clay. *Here Comes Everybody: The Power of Organizing without Organizations.* New York: Penguin, 2008.

Shulman, Max. *Max Shulman's Large Economy Size.* New York: Doubleday & Company, 1948.

Siedman, Robert. "BBC America Starts 2010 With Best Ever Night and Day." *TV By the Numbers.* 06 Jan 2009. http://tvbythenumbers.zap2it.com/2010/01/06/bbc-america-starts-2010-with-best-ever-night-and-day/37825 (accessed 11 Apr 2011).

Simmons, Bruce. "'Supernatural' Episode 'The French Mistake' Review and Misha's Tweet." *Brusimm.* 26 Feb 2011. http://brusimm.com/2011/02/26/supernatural-episode-the-french-mistake-review-mishas-tweet/ (accessed 14 Aug 2011).

SISAPSIS. "Sitting Dance." *YouTube.* 16 Feb 2008. http://www.youtube.com/watch?v=fnFiqRiZh1s (accessed 21 Mar 2011).

Smith, Adam, writer. "Flesh and Stone." *Doctor Who.* London: BBC, 2010.

Smith, Aaron. "Why Americans Use Social Media." *Pew Research Center.* 15 Nov 2011. http://www.pewinternet.org/Reports/2011/Why-Americans-Use-Social-Media/Main-report.aspx (accessed 02 Dec 2011).

Sobchack, Vivian. *The Persistence of History: Cinema, Television and the Modern Event.* New York: Routledge, 1995.

———. *Screening Space: The American Science Fiction Film.* New Brunswick, NJ: Rutgers University Press, 1987.

Somers, Margaret. "The Narrative Constitution of Identity: A Relationship and Network Approach." *Theory and Society* 23 (1994): 605–49.

Spigel, Lynn. "Introduction." In *Television after TV: Essays on a Medium in Transition,* edited by Lynn Spigel and Jan Olsson, 1–39. Durham, NC: Duke University Press, 2004.

Starner, Thad, Steve Mann, Bradley Rhodes, Jeffrey Levine, Jennifer Healey, Dana Kirsch, Roz Picard, and Alex Pentland. "Augmented Reality Through Wearable Computing." *Presence: Teleoperators and Virtual Environments* 6, no. 4 (1997): http://www.cc.gatech.edu/fac/Thad.Starner/p/journal/augmented-reality-through-wearable-computing.pdf (accessed 14 Mar 2011).

Sterling, Bruce. *Shaping Things.* Cambridge, MA: The MIT Press, 2005.

Strangelove, Michael. *Watching YouTube: Extraordinary Videos by Ordinary People.* Toronto, ON: University of Toronto Press, 2010.

Stvilia, Besiki, Michael B. Twidale, Linda C. Smith, and Les Gasser. "Information Quality Work Organization in Wikipedia." *Journal of the American Society for Information Science and Technology* 59, no. 6 (2008): 983–1001.

Suddendorf, Thomas, and Michael Corballis. "Mental Time Travel and the Evolution of the Human Mind." *Genetic, Social, and General Psychology Monographs* 123, no. 2 (1997): 133–67.

Tapscott, Don, and Anthony Williams. *Wikinomics: How Mass Collaboration Changes Everything.* New York: Penguin, 2006.

"Teen Leisure Habits: TV Still On Top." *Media Literacy Clearinghouse.* 28 Mar 2011. http://www.frankwbaker.com/mediause.htm (accessed 01 Mar 2012).

"The Eleventh Hour." *Tardis.wikia.* http://tardis.wikia.com/wiki/The_Eleventh_Hour#Synopsis (accessed 02 Apr 2010).

Thorburn, David. "Television as an Aesthetic Medium." *Critical Studies in Mass Communication* 4, no. 2 (1987): 61–73.

Thorburn, David, and Henry Jenkins. "Introduction: Toward an Aesthetics of Transition." In *Rethinking Media Change: The Aesthetics of Transition,* edited by David Thorburn and Henry Jenkins, 1–16. Cambridge, MA: The MIT Press, 2004.

Toplin, Robert. *History by Hollywood: The Use and Abuse of the American Past.* Champaign: University of Illinois Press, 1996.

Tryon, Chuck. *Reinventing Cinema: Movies in the Age of Media Convergence.* New Brunswick, NJ: Rutgers University Press, 2009.

Tulloch, John, and Manuel Alvarado, *Doctor Who: The Unfolding Text.* New York: St. Martins Press, 1984.

Turkle, Sherry. *Alone Together.* New York: Basic Books, 2011.

Uricchio, William. "TV as Time Machine: Television's Changing Heterochronic Regimes and the Production of History." In *Relocating Television: Television in the Digital Context,* edited by Jostein Gripsrud, 27–40. London: Routledge, 2010.

van den Hoven, Elise, and Berry Eggen. "Informing Augmented Memory System Design through Autobiographical Memory Theory." *Personal and Ubiquitous Computing* 12 (2008): 433–43.

van Dijk, Jan. *Network Society: Social Aspects of the New Media.* 2nd ed. London: Sage Publications, 2005.

van Dijk, José. *Mediated Memories in the Digital Age.* Stanford, CA: Stanford University Press, 2007.

van Kokswijk, Jacob. *Digital Ego: Social and Legal Aspects of Virtual Identity.* 2007. Rev. ed. Delft, The Netherlands: Eburon Academic Publishers, 2008.

Vermeule, Blakey. *Why Do We Care about Literary Characters?* Baltimore, MD: The Johns Hopkins University Press, 2010.

Voigts-Virchow, Eckart. "History: The Sitcom, England: The Theme Park—*Blackadder*'s Retrovisions as Historiographic Meta-TV." In *Narrative Strategies in Television Series,* edited by Gaby Allrath and Marion Gymnich, 211–28. Houndmills, UK: Palgrave Macmillan, 2006.

Walker, Jill. "Distributed Narratives: Telling Stories across Networks." 2007. http://jilltxt.net/txt/Walker-AoIR-3500words.pdf (accessed 01 May 2010).

Walsh, Richard. "Fabula and Fictionality in Narrative Theory." *Style* 35, no. 4 (2001): 592–606.

———. *The Rhetoric of Fictionality: Narrative Theory and the Idea of Fiction.* Columbus, OH: Ohio State University, 2007.

Welch, Kathleen. *Electric Rhetoric: Classical Rhetoric, Oralism, and a New Literacy.* Cambridge, MA: The MIT Press, 1999.

White, Cindy. "Executive Producers Damon Lindelof and Carlton Cuse Promise That They've Found the Plot Twists That Will Bring *Lost*'s Viewers Back." *Sci-Fi.com.* 22 Jan 2007. http://web.archive.org/web/

20070318060759/http://www.scifi.com/sfw/interviews/sfw14769.html (accessed 02 Mar 2011).

White, Hayden. *The Content of the Form: Narrative Discourse and Historical Representation*. Baltimore, MD: The Johns Hopkins University Press, 1990.

———. *Metahistory: The Historical Imagination in Nineteenth-Century Europe*. Baltimore, MD: The Johns Hopkins University Press, 1973.

White, Michele. "Television and Internet Differences by Design: Rendering Liveness, Presence, and Lived Space." *Convergence: The International Journal of Research into New Media Technologies* 12, no. 3 (2006): 341–55.

Williams, Raymond. *Cultural and Materialism*. 1980. New York: Verso, 2005.

———. *Culture and Society, 1780–1950*. New York: Columbia University Press, 1958.

———. *Keywords: A Vocabulary of Culture and Society*. New York: Oxford University Press, 1976.

———. *Television: Technology and Cultural Form*. 1974. New York: Routledge, 2003.

Williams, Siôn, and John Keady. "Editorial: The Narrative Voice of People with Dementia." *Dementia* 5, no. 2 (2006): 163–66.

Wood, Tat. *About Time: The Unauthorized Guide to Doctor Who, 1985–1989, Seasons 22 to 26, The TV Movie*. Des Moines, IA: Mad Norwegian Press, 2007.

Wood, Tat, and Lawrence Miles. *About Time: The Unauthorized Guide to Doctor Who, 1963–1966, Seasons 1 to 3*. Des Moines, IA: Mad Norwegian Press, 2006.

———. *About Time: The Unauthorized Guide to Doctor Who 1966–1969, Seasons 4 to 6*. Des Moines, IA: Mad Norwegian Press, 2006.

Wortham, Janna. "Friendster to Erase Early Posts and Old Photos." *New York Times*. 27 Apr 2011. http://www.nytimes.com/2011/04/27/technology/27friendster.html (accessed 27 Apr 2011).

Yu, Charles. *How to Live Safely in a Science Fictional Universe*. New York: Pantheon Books, 2010.

Zalinger, Jason. "Gmail as Storyworld: How Technology Shapes Your Life Narrative." PhD diss., Rensselaer Polytechnic Institute, 2011.

Zittrain, Jonathan. *The Future of the Internet and How to Stop It*. New Haven, CT: Yale University Press, 2008.

Index

Index

C

D

H

I

J

K

L

O

P

Q

R

S

T

www.ingramcontent.com/pod-product-compliance
Ingram Content Group UK Ltd.
Pitfield, Milton Keynes, MK11 3LW, UK
UKHW021851100926
8556IPUK00020B/1178